Vegetable and Fruits: Farming and Gardening

Vegetable and Fruits: Farming and Gardening

Edited by
Braden Bowen

Larsen & Keller
www.larsen-keller.com

Vegetable and Fruits: Farming and Gardening
Edited by Braden Bowen
ISBN: 978-1-63549-282-8 (Hardback)

© 2017 Larsen & Keller

▤ Larsen & Keller

Published by Larsen and Keller Education,
5 Penn Plaza,
19th Floor,
New York, NY 10001, USA

Cataloging-in-Publication Data

Vegetable and fruits : farming and gardening / edited by Braden Bowen.
    p. cm.
Includes bibliographical references and index.
ISBN 978-1-63549-282-8
1. Vegetables. 2. Fruit. 3. Vegetable gardening. 4. Fruit-culture.
I. Bowen, Braden.
SB321 .V44 2017
635--dc23

The publisher's policy is to use permanent paper from mills that operate a sustainable forestry policy. Furthermore, the publisher ensures that the text paper and cover boards used have met acceptable environmental accreditation standards.

Printed and bound in the United States of America.

For more information regarding Larsen and Keller Education and its products, please visit the publisher's website www.larsen-keller.com

Table of Contents

Preface

This book is a compilation of chapters that discuss the most vital concepts in the field of vegetable and fruit farming and gardening. It discusses in detail the various theories and techniques related to this field in depth. Vegetable and fruit farming refers to the practice of growing vegetables and fruits for human consumption. It uses methods like raised bed, aquaponics, etc. to improve yield and produce. The topics included in this book are of utmost significance and are bound to provide incredible insights to readers. The textbook aims to serve as a resource guide for students and facilitate the study of the discipline.

A short introduction to every chapter is written below to provide an overview of the content of the book:

Chapter 1 - Vegetable farming is the process of growing vegetables for the purpose of consumption. Vegetable farming was initially performed manually but as times have evolved and mechanization implemented, the processes are performed by machines. This chapter will provide an integrated understanding of farming and gardening; **Chapter 2 -** A farm is an area that is particularly used in agricultural processes. The basic objective is the production of food crops and food production. Some of the topics discussed in this section are herb farm, square foot gardening, forest gardening, organic horticulture, kitchen garden and slow gardening. This section is an overview of the subject matter incorporating all the major aspects of farm and gardening; **Chapter 3 -** An orchard is the planting of trees for food production. They comprise of fruit producing trees or nut producing trees; they also feature large gardens that serve an aesthetic purpose. The chapter will not only provide an outline, it will also delve deep into the topics related to it; **Chapter 4 -** Hydroponics is the method of growing plant without soil; instead of using soil, mineral nutrient solutions are used in a water solvent. Aquaponics, hydroponic dosers and historical hydroculture are some of the aspects of hydroponics that have been elucidated in the following section; **Chapter 5 -** Climate-friendly gardening helps in the reduction of greenhouse gasses and aids in the reduction of global warming. The alternative ways of environment friendly gardening are agroforestry, permaculture, wildlife garden, forest gardening etc. The aspects elucidated in this chapter are of vital importance and provide a better understanding of environmental friendly gardening; **Chapter 6 -** The history of ornamental gardening is a way of displaying beauty through art and nature. It is put on display either for private status or for national pride. The section helps the readers in understanding the evolution of gardening.

I extend my sincere thanks to the publisher for considering me worthy of this task. Finally, I thank my family for being a source of support and help.

Editor

Introduction to Vegetable Farming and Gardening

Vegetable farming is the process of growing vegetables for the purpose of consumption. Vegetable farming was initially performed manually but as times have evolved and mechanization implemented, the processes are performed by machines. This chapter will provide an integrated understanding of farming and gardening.

Vegetable Farming

Vegetable farming is the growing of vegetables for human consumption. The practice probably started in several parts of the world over ten thousand years ago, with families growing vegetables for their own consumption or to trade locally. At first manual labour was used but in time livestock were domesticated and the ground could be turned by the plough. More recently, mechanisation has revolutionised vegetable farming with nearly all processes being able to be performed by machine. Specialist producers grow the particular crops that do well in their locality. New methods such as aquaponics, raised beds and cultivation under glass are used. Marketing can be done locally in farmer's markets, traditional markets or pick-your-own operations, or farmers can contract their whole crops to wholesalers, canners or retailers.

History

Originally, vegetables were collected from the wild by hunter-gatherers and entered cultivation in several parts of the world, probably during the period 10,000 BC to 7,000 BC, when a new agricultural way of life developed. At first, plants which grew locally would have been cultivated, but as time went on, trade brought exotic crops from elsewhere to add to domestic types. Nowadays, most vegetables are grown all over the world as climate permits.

Traditionally it was done in the soil in small rows or blocks, often primarily for consumption on the farm, with the excess sold in nearby towns. Later, farms on the edge of large communities could specialize in vegetable production, with the short distance allowing the farmer to get his produce to market while still fresh. The three sisters method used by Native Americans (specifically the Haudenosaunee/Iroquois) grew squash, beans and corn together so that the plants enhanced each other's growth. Planting in long rows allows machinery to cultivate the fields, increasing efficiency and output; however, the diversity of vegetable crops requires a number of techniques to be used to optimize the growth of each type of plant. Some farms, therefore, specialize in one vegetable; others grow a large variety. Due to the needs to market vegetables while fresh, vegetable gardening has high labor demands. Some farms avoid this by running u-pick operations where the customers pick their own produce. The development of ripening technologies and refrigeration has reduced the problems with getting produce to market in good condition.

Different Growing Methods

Over the past 100 years a new technique has emerged—raised bed gardening, which has increased yields from small plots of soil without the need for commercial, energy-intensive fertilizers. Modern hydroponic farming produces very high yields in greenhouses without using any soil.

Marketing

Several economic models exist for vegetable farms: farms may grow large quantities of a few vegetables and sell them in bulk to major markets or middlemen, which requires large growing operations; farms may produce for local customers, which requires a larger distribution effort; farms may produce a variety of vegetables for sale through on-farm stalls, local farmer's markets, u-pick operations. This is quite different from commodity farm products like wheat and maize which do not have the ripeness problems and are sold off in bulk to the local granary. Large cities often have a central produce market which handles vegetables in a commodity-like manner, and manages distribution to most supermarkets and restaurants.

In America, vegetable farms are in some regions known as truck farms; "truck" is a noun for which its more common meaning overshadows its historically separate use as a term for "vegetables grown for market". Such farms are sometimes called muck farms, after the dark black soil in which vegetables grow well.

Common Vegetable Crops

Vegetables which are farmed include:

- Fabaceae (pea family): peas, beans, lentils
- Solanaceae (nightshade family): tomatoes, eggplants, bell peppers, potatoes
- Brassicaceae (mustard family): cauliflower, cabbage, brussels sprouts, broccoli
- Allium family: onions, garlic, leek, shallot, chives
- Carrots (Apiaceae)

- Lettuce (Asteraceae)
- cucurbit family of plants including melon, cantaloupe, cucumber, calabash, squash, and pumpkin
- Sweet corn

Gardening

Part of a parterre in an English garden

Gardening is the practice of growing and cultivating plants as part of horticulture. In gardens, ornamental plants are often grown for their flowers, foliage, or overall appearance; useful plants, such as root vegetables, leaf vegetables, fruits, and herbs, are grown for consumption, for use as dyes, or for medicinal or cosmetic use. Gardening is considered to be a relaxing activity for many people.

Gardening ranges in scale from fruit orchards, to long boulevard plantings with one or more different types of shrubs, trees, and herbaceous plants, to residential yards including lawns and foundation plantings, to plants in large or small containers grown inside or outside. Gardening may be very specialized, with only one type of plant grown, or involve a large number of different plants in mixed plantings. It involves an active participation in the growing of plants, and tends to be labor-intensive, which differentiates it from farming or forestry.

Ancient Times

Forest gardening, a forest-based food production system, is the world's oldest form of gardening. Forest gardens originated in prehistoric times along jungle-clad river banks and in the wet foothills of monsoon regions. In the gradual process of families improving their immediate environment, useful tree and vine species were identified, protected and improved while undesirable species were eliminated. Eventually foreign species were also selected and incorporated into the gardens.

After the emergence of the first civilizations, wealthy individuals began to create gardens for aesthetic purposes. Egyptian tomb paintings from around 1500 BC provide some of the earliest physical evidence of ornamental horticulture and landscape design; they depict lotus ponds surrounded by symmetrical rows of acacias and palms. A notable example of ancient ornamental gardens were the Hanging Gardens of Babylon—one of the Seven Wonders of the Ancient World —while ancient Rome had dozens of gardens.

Wealthy ancient Egyptians used gardens for providing shade. Egyptians associated trees and gardens with gods as they believed that their deities were pleased by gardens. Gardens in ancient Egypt were often surrounded by walls with trees planted in rows. Among the most popular species planted were date palms, sycamores, fir trees, nut trees, and willows. These gardens were a sign of higher socioeconomic status. In addition, wealthy ancient Egyptians grew vineyards, as wine was a sign of the higher social classes. Roses, poppies, daisies and irises could all also be found in the gardens of the Egyptians.

Assyria was also renowned for its beautiful gardens. These tended to be wide and large, some of them used for hunting game—rather like a game reserve today—and others as leisure gardens. Cypresses and palms were some of the most frequently planted types of trees.

Ancient Roman gardens were laid out with hedges and vines and contained a wide variety of flowers—acanthus, cornflowers, crocus, cyclamen, hyacinth, iris, ivy, lavender, lilies, myrtle, narcissus, poppy, rosemary and violets—as well as statues and sculptures. Flower beds were popular in the courtyards of rich Romans.

A gardener at work, 1607

The Middle Ages

The Middle Age represented a period of decline in gardens for aesthetic purposes, with regard to gardening. After the fall of Rome, gardening was done for the purpose of growing medicinal herbs and/or decorating church altars. Monasteries carried on a tradition of garden design and intense horticultural techniques during the medieval period in Europe. Generally, monastic garden types consisted of kitchen gardens, infirmary gardens, cemetery orchards, cloister garths and vineyards. Individual monasteries might also have had a "green court", a plot of grass and trees where horses could graze, as well as a cellarer's garden or private gardens for obedientiaries, monks who held specific posts within the monastery.

Islamic gardens were built after the model of Persian gardens and they were usually enclosed by walls and divided in 4 by watercourses. Commonly, the center of the garden would have a pool or pavilion. Specific to the Islamic gardens are the mosaics and glazed tiles used to decorate the rills and fountains that were built in these gardens.

By the late 13th century, rich Europeans began to grow gardens for leisure and for medicinal herbs and vegetables. They surrounded the gardens by walls to protect them from animals and to provide seclusion. During the next two centuries, Europeans started planting lawns and raising flowerbeds and trellises of roses. Fruit trees were common in these gardens and also in some, there were turf seats. At the same time, the gardens in the monasteries were a place to grow flowers and medicinal herbs but they were also a space where the monks could enjoy nature and relax.

The gardens in the 16th and 17th century were symmetric, proportioned and balanced with a more classical appearance. Most of these gardens were built around a central axis and they were divided into different parts by hedges. Commonly, gardens had flowerbeds laid out in squares and separated by gravel paths.

Gardens in Renaissance were adorned with sculptures, topiary and fountains. In the 17th century, knot gardens became popular along with the hedge mazes. By this time, Europeans started planting new flowers such as tulips, marigolds and sunflowers.

Cottage Gardens

Cottage gardens, which emerged in Elizabethan times, appear to have originated as a local source for herbs and fruits. One theory is that they arose out of the Black Death of the 1340s, when the death of so many laborers made land available for small cottages with personal gardens. According to the late 19th-century legend of origin, these gardens were originally created by the workers that lived in the cottages of the villages, to provide them with food and herbs, with flowers planted among them for decoration. Farm workers were provided with cottages that had architectural quality set in a small garden—about an acre—where they could grow food and keep pigs and chickens.

Authentic gardens of the yeoman cottager would have included a beehive and livestock, and frequently a pig and sty, along with a well. The peasant cottager of medieval times was more interested in meat than flowers, with herbs grown for medicinal use rather than for their beauty. By Elizabethan times there was more prosperity, and thus more room to grow flowers. Even the early cottage garden flowers typically had their practical use—violets were spread on the floor (for their pleasant scent and keeping out vermin); calendulas and primroses were both attractive and used in cooking. Others, such as sweet william and hollyhocks, were grown entirely for their beauty.

18th Century

In the 18th century, gardens were laid out more naturally, without any walls. This style of smooth undulating grass, which would run straight to the house, clumps, belts and scattering of trees and his serpentine lakes formed by invisibly damming small rivers, were a new style within the English landscape, a "gardenless" form of landscape gardening, which swept away almost all the remnants of previous formally patterned styles. The English garden usually included a lake, lawns set against groves of trees, and often contained shrubberies, grottoes, pavilions, bridges and follies such as mock temples, Gothic ruins, bridges, and other picturesque architecture, designed to recreate an idyllic pastoral landscape. This new style emerged in England in the early 18th century, and spread across Europe, replacing the more formal, symmetrical Garden à la française of the 17th century as the principal gardening style of Europe. The English garden presented an idealized view of nature. They were often inspired by paintings of landscapes by Claude Lorraine and Nicolas Poussin, and some were Influenced by the classic Chinese gardens of the East, which had recently been described by European travelers. The work of Lancelot 'Capability' Brown was particularly influential. Also, in 1804 the Horticultural Society was formed. Gardens of the 19th century contained plants such as the monkey puzzle or Chile pine. This is also the time when the so-called "gardenesque" style of gardens evolved. These gardens displayed a wide variety of flowers in a rather small space. Rock gardens increased in popularity in the 19th century.

Types

Conservatory of Flowers in Golden Gate Park, San Francisco

Hanging baskets in Thornbury, South Gloucestershire

Residential gardening takes place near the home, in a space referred to as the garden. Although a garden typically is located on the land near a residence, it may also be located on a roof, in an atrium, on a balcony, in a windowbox, or on a patio or vivarium.

Gardening also takes place in non-residential green areas, such as parks, public or semi-public gardens (botanical gardens or zoological gardens), amusement and amusement parks, along transportation corridors, and around tourist attractions and garden hotels. In these situations, a staff of gardeners or groundskeepers maintains the gardens.

- Indoor gardening is concerned with the growing of houseplants within a residence or building, in a conservatory, or in a greenhouse. Indoor gardens are sometimes incorporated as part of air conditioning or heating systems. Indoor gardening extends the growing season in the fall and spring and can be used for winter gardening.

- Native plant gardening is concerned with the use of native plants with or without the intent of creating wildlife habitat. The goal is to create a garden in harmony with, and adapted to a given area. This type of gardening typically reduces water usage, maintenance, and fertilization costs, while increasing native faunal interest.

- Water gardening is concerned with growing plants adapted to pools and ponds. Bog gardens are also considered a type of water garden. These all require special conditions and considerations. A simple water garden may consist solely of a tub containing the water and plant(s). In aquascaping, a garden is created within an aquarium tank.

- Container gardening is concerned with growing plants in any type of container either indoors or outdoors. Common containers are pots, hanging baskets, and planters. Container gardening is usually used in atriums and on balconies, patios, and roof tops.

- Hügelkultur is concerned with growing plants on piles of rotting wood, as a form of raised bed gardening and composting in situ. An English loanword from German, it means "mound garden." Toby Hemenway, noted Permaculture author and teacher, considers wood buried in trenches to also be a form of hugelkultur referred to as a dead wood swale. Hugelkultur is practiced by Sepp Holzer as a method of forest gardening and agroforestry, and by Geoff Lawton as a method of dryland farming and desert greening. When used as a method of disposing of large volumes of waste wood and woody debris, hugelkultur accomplishes carbon sequestration. It is also a form of xeriscaping.

- Community gardening is a social activity in which an area of land is gardened by a group of people, providing access to fresh produce and plants as well as access to satisfying labor, neighborhood improvement, sense of community and connection to the environment. Community gardens are typically owned in trust by local governments or nonprofits.

- Garden sharing partners landowners with gardeners in need of land. These shared gardens, typically front or back yards, are usually used to produce food that is divided between the two parties.

- Organic gardening uses natural, sustainable methods, fertilizers and pesticides to grow non-genetically modified crops.

Garden Features and Accessories

There is a wide range of features and accessories available in the market for both the professional gardener and the amateur to exercise their creativity. These are used to add decoration or functionality, and may be made from a wide range of materials such as copper, stone, wood, bamboo, stainless steel, clay, stained glass, concrete, or iron. Examples include trellis, arbors, statues, benches, water fountains, urns, bird baths and feeders, and garden lighting such as candle lanterns and oil lamps. The use of these items can be part of the expression of a gardener's gardening personality.

Gardening Departments and Centers

Gardening departments and centers mainly sell plants, sundries, and garden accessories, but in recent times,[when?] many now stock outdoor leisure products as diverse as spas, furniture, and barbecues. Many garden centers now include food halls, and sections for clothing, gifts, pets, and power tools. There are also a number of online garden centers that now deliver directly to customers' doors.

Comparison with Farming

Gardening for beauty is likely nearly as old as farming for food, however for most of history for the majority of people there was no real distinction since the need for food and other useful product trumped other concerns. Small-scale, subsistence agriculture (called hoe-farming) is largely indistinguishable from gardening. A patch of potatoes grown by a Peruvian peasant or an Irish smallholder for personal use could be described as either a garden or a farm. Gardening for average people evolved as a separate discipline, more concerned with aesthetics and recreation, under the influence of the pleasure gardens of the wealthy. Meanwhile, farming has evolved (in developed countries) in the direction of commercialization, economics of scale, and monocropping.

Hand gardening tools

In respect to its food producing purpose, gardening is distinguished from farming chiefly by scale and intent. Farming occurs on a larger scale, and with the production of salable goods as a major motivation. Gardening is done on a smaller scale, primarily for pleasure and to produce goods for the gardener's own family or community. There is some overlap between the terms, particularly in that some moderate-sized vegetable growing concerns, often called market gardening, can fit in either category.

Planting in a garden

The key distinction between gardening and farming is essentially one of scale; gardening can be a hobby or an income supplement, but farming is generally understood as a full-time or commercial activity, usually involving more land and quite different practices. One distinction is that gardening is labor-intensive and employs very little infrastructural capital, sometimes no more than a few tools, e.g. a spade, hoe, basket and watering can. By contrast, larger-scale farming often involves irrigation systems, chemical fertilizers and harvesters or at least ladders, e.g. to reach up into fruit trees. However, this distinction is becoming blurred with the increasing use of power tools in even small gardens.

In part because of labor intensity and aesthetic motivations, gardening is very often much more productive per unit of land than farming. In the Soviet Union, half the food supply came from small peasants' garden plots on the huge government-run collective farms, although they were tiny patches of land. Some argue this as evidence of the superiority of capitalism, since the peasants were generally able to sell their produce. Others consider it to be evidence of a tragedy of the commons, since the large collective plots were often neglected, or fertilizers or water redirected to the private gardens.

The term precision agriculture is sometimes used to describe gardening using intermediate technology (more than tools, less than harvesters), especially of organic varieties. Gardening is effectively scaled up to feed entire villages of over 100 people from specialized plots. A variant is the community garden which offers plots to urban dwellers.

Gardens as Art

Garden at the Schultenhof in Mettingen, North Rhine-Westphalia, Germany

Garden design is considered to be an art in most cultures, distinguished from gardening, which generally means garden maintenance. Garden design can include different themes such as perennial, butterfly, wildlife, Japanese, water, tropical, or shade gardens. In Japan, Samurai and Zen monks were often required to build decorative gardens or practice related skills like flower arrangement known as ikebana. In 18th-century Europe, country estates were refashioned by landscape gardeners into formal gardens or landscaped park lands, such as at Versailles, France, or Stowe, England. Today, landscape architects and garden designers continue to produce artistically creative designs for private garden spaces. In the USA, professional landscape designers are certified by the Association of Professional Landscape Designers.

Social Aspects

People can express their political or social views in gardens, intentionally or not. The lawn vs. garden issue is played out in urban planning as the debate over the "land ethic" that is to determine urban land use and whether hyper hygienist bylaws (e.g. weed control) should apply, or whether land should generally be allowed to exist in its natural wild state. In a famous Canadian Charter of Rights case, "Sandra Bell vs. City of Toronto", 1997, the right to cultivate all native species, even most varieties deemed noxious or allergenic, was upheld as part of the right of free expression.

Community gardening comprises a wide variety of approaches to sharing land and gardens.

People often surround their house and garden with a hedge. Common hedge plants are privet, hawthorn, beech, yew, leyland cypress, hemlock, arborvitae, barberry, box, holly, oleander, forsythia and lavender. The idea of open gardens without hedges may be distasteful to those who enjoy privacy. The Slow Food movement has sought in some countries to add an edible school yard and garden classrooms to schools, e.g. in Fergus, Ontario, where these were added to a public school to augment the kitchen classroom. Garden sharing, where urban landowners allow gardeners to grow on their property in exchange for a share of the harvest, is associated with the desire to control the quality of one's food, and reconnect with soil and community.

In US and British usage, the production of ornamental plantings around buildings is called landscaping, landscape maintenance or grounds keeping, while international usage uses the term gardening for these same activities.

Also gaining popularity is the concept of "Green Gardening" which involves growing plants using organic fertilizers and pesticides so that the gardening process - or the flowers and fruits produced thereby - doesn't adversely affect the environment or people's health in any manner.

Garden Pests

Garden pests are generally plants, fungi, or animals (frequently insects) that engages in activity that the gardener considers undesirable. It may crowd out desirable plants, disturb soil, stunt the growth of young seedlings, steal or damage fruit, or otherwise kill plants, hamper their growth, damage their appearance, or reduce the quality of the edible or ornamental portions of the plant. Aphids, spider mites, slugs, snails, ants, birds, and even cats are commonly considered to be garden pests.

Because gardeners may have different goals, organisms considered "garden pests" vary from gardener to gardener. Tropaeolum speciosum, for example, may be considered a desirable and ornamental garden plant, or it may be considered a pest if it seeds and starts to grow where it is not wanted. As another example, in lawns, moss can become dominant and be impossible to eradicate. In some lawns, lichens, especially very damp lawn lichens such as Peltigera lactucfolia and P. membranacea, can become difficult to control and be considered pests.

Garden Pest Control

There are many ways by which unwanted pests are removed from a garden. The techniques vary depending on the pest, the gardener's goals, and the gardener's philosophy. For example, snails may be dealt with through the use of a chemical pesticide, an organic pesticide, hand-picking, barriers, or simply growing snail-resistant plants.

Pest control is often done through the use of pesticides, which may be either organic or artificially synthesized. Pesticides may affect the ecology of a garden due to their effects on the populations of both target and non-target species. For example, unintended exposure to some neonicotinoid pesticides has been proposed as a factor in the recent decline in honey bee populations. A mole vibrator can deter mole activity in a garden.

Other means of control include the removal of infected plants, using fertilizers and biostimulants to improve the health and vigour of plants so they better resist attack, practising crop rotation to prevent pest build-up, using companion planting, and practising good garden hygiene, such as disinfecting tools and clearing debris and weeds which may harbour pests.

Raised-bed Gardening

Raised garden bed of lettuce, tomatoes, basil, marigolds, zinnias, garlic chives, zucchini.

Raised-bed gardening is a form of gardening in which the soil is formed in three-to-four-foot-wide (1.0–1.2 m) beds, which can be of any length or shape. The soil is raised above the surrounding soil (approximately six inches to waist-high), is sometimes enclosed by a frame generally made of wood, rock, or concrete blocks, and may be enriched with compost. The vegetable plants are spaced in geometric patterns, much closer together than in conventional row gardening. The spacing is such that when the vegetables are fully grown, their leaves just barely touch each other, creating a micro-climate in which weed growth is suppressed and moisture is conserved. Raised beds produce a variety of benefits: they extend the planting season, they can reduce weeds if designed and planted properly, and they reduce the need to use poor native soil. Since the gardener does not walk on the raised beds, the soil is not compacted and the roots have an easier time growing. The close plant spacing and the use of compost generally result in higher yields with raised beds in comparison to conventional row gardening. Waist-high raised beds enable the elderly and physically disabled to grow vegetables without having to bend over to tend them.

Raised bed gardening

Picardo Farm, Wedgwood neighborhood, Seattle, Washington: A community allotment garden with raised beds for the physically disabled.

Overview

Raised beds lend themselves to the development of complex agriculture systems that utilize many of the principles and methods of permaculture. They can be used effectively to control erosion and recycle and conserve water and nutrients by building them along contour lines on slopes. This also makes more space available for intensive crop production. They can be created over large areas with the use of several commonly available tractor-drawn implements and efficiently maintained, planted and harvested using hand tools.

This form of gardening is compatible with square foot gardening and companion planting.

Circular raised beds with a path to the center (a slice of the circle cut out) are called keyhole gardens. Often the center has a chimney of sorts built with sticks and then lined with feedbags or grasses that allows water placed at the center to flow out into the soil and reach the plants' roots.

Materials and Construction

Vegetable garden bed construction materials should be chosen carefully. Some concerns exist regarding the use of pressure-treated timber. Pine that was treated using chromated copper arsenate or CCA, a toxic chemical mix for preserving timber that may leach chemicals into the soil which in turn can be drawn up into the plants, is a concern for vegetable growers, where part or all of the plant is eaten. If using timber to raise the garden bed, ensure that it is an untreated hardwood to prevent the risk of chemicals leaching into the soil. A common approach is to use timber sleepers joined with steel rods to hold them together. Another approach is to use concrete blocks, although less aesthetically pleasing, they are inexpensive to source and easy to use. On the market are also prefab raised garden bed solutions which are made from long lasting polyethylene that is UV stabilized and food grade so it will not leach undesirable chemicals into the soil or deteriorate in the elements. A double skinned wall provides an air pocket of insulation that minimizes the temperature fluctuations and drying out of the soil in the garden bed. Sometimes raised bed gardens are covered with clear plastic to protect the crops from wind and strong rains. Pre-manufactured raised bed gardening boxes also exist. There are variants of wood, metal, stone and plastic. Each material type has advantages and disadvantages.

References

- Douglas John McConnell (2003). The Forest Farms of Kandy: And Other Gardens of Complete Design. p. 1. ISBN 9780754609582.

- Ryrie, Charlie (2004). The Cottage Garden: How to Plan and Plant a Garden That Grows Itself. Collins & Brown. p. 7. ISBN 1-84340-216-5.

- Scott-James, Anne; Osbert Lancaster (2004). The Pleasure Garden: An Illustrated History of British Gardening. Frances Lincoln Publishers. p. 80. ISBN 978-0-7112-2360-8.

- A Biographical Dictionary of British Architects, 1600–1840, Howard Colvin, Yale University Press, 2008 ISBN 0-300-12508-9, p 659

- Lloyd, Christopher; Richard Bird (1999). The Cottage Garden. Jacqui Hurst. Dorling Kindersley. pp. 6–9. ISBN 978-0-7513-0702-3.

- Boults, Elizabeth and Chip Sullivan (2010). Illustrated History of Landscape Design. John Wiley and Sons. p. 175. ISBN 0-470-28933-3.

- Hemenway, Toby (2009). Gaia's Garden: A Guide to Home-Scale Permaculture. Chelsea Green Publishing. pp. 84-85. ISBN 978-1-60358-029-8.

- Hughes, Megan McConnell (2010). Better Homes & Gardens Vegetable, Fruit & Herb Gardening. Wiley. pp. 68–69. Retrieved March 2, 2012. ISBN 978-0-470-63856-9

- Nones, Raymond (2010). Raised-Bed Vegetable Gardening Made Simple. Countryman Press. Retrieved March 2, 2012. ISBN 978-0-88150-896-3

- Whiting, David E. (1991). The desert shall blossom: a comprehensive guide to vegetable gardening in the Mountain West. Horizon. pp. 41–42. Retrieved March 2, 2012. ISBN 0-88290-418-3

- Kemery, Ricky (January 29, 2012). "Unlock your creativity with keyhole garden". The Journal Gazette (Fort Wayne, IN). Retrieved March 6, 2012.

- Lively, Ruth. "Does Pressure-Treated Wood Belong in Your Garden?". Fine Gardening Magazine. Retrieved March 6, 2012.

Farm and Garden: An Overview

A farm is an area that is particularly used in agricultural processes. The basic objective is the production of food crops and food production. Some of the topics discussed in this section are herb farm, square foot gardening, forest gardening, organic horticulture, kitchen garden and slow gardening. This section is an overview of the subject matter incorporating all the major aspects of farm and gardening.

Farm

Farmland in the USA. The round fields are due to the use of center pivot irrigation

A farm is an area of land that is devoted primarily to agricultural processes with the primary objective of producing food and other crops; it is the basic facility in food production. The name is used for specialised units such as arable farms, vegetable farms, fruit farms, dairy, pig and poultry farms, and land used for the production of natural fibres, biofuel and other commodities. It includes ranches, feedlots, orchards, plantations and estates, smallholdings and hobby farms, and includes the farmhouse and agricultural buildings as well as the land. In modern times the term has been extended so as to include such industrial operations as wind farms and fish farms, both of which can operate on land or sea.

Farming originated independently in different parts of the world, as hunter gatherer societies transitioned to food production rather than, food capture. It may have started about 12,000 years ago with the domestication of livestock in the Fertile Crescent in western Asia, soon to be followed by the cultivation of crops. Modern units tend to specialise in the crops or livestock best suited to the region, with their finished products being sold for the retail market or for further processing, with farm products being traded around the world.

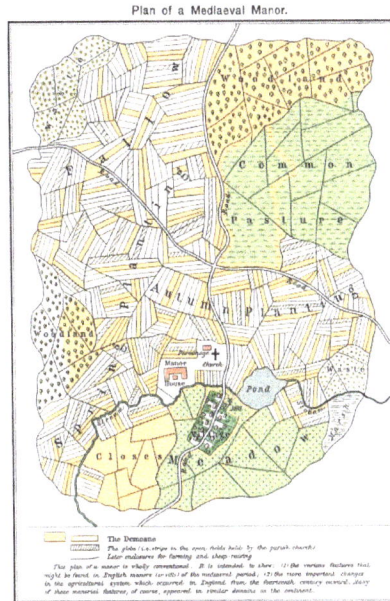

Typical plan of a mediaeval English manor, showing the use of field strips

Modern farms in developed countries are highly mechanized. In the United States, livestock may be raised on rangeland and finished in feedlots and the mechanization of crop production has brought about a great decrease in the number of agricultural workers needed. In Europe, traditional family farms are giving way to larger production units. In Australia, some farms are very large because the land is unable to support a high stocking density of livestock because of climatic conditions. In less developed countries, small farms are the norm, and the majority of rural residents are subsistence farmers, feeding their families and selling any surplus products in the local market.

Etymology

A farmer harvesting crops with mule-drawn wagon, 1920s, Iowa, USA

The word in the sense of an agricultural land-holding derives from the verb "to farm" a revenue source, whether taxes, customs, rents of a group of manors or simply to hold an individual manor by the feudal land tenure of "fee farm". The word is from the medieval Latin noun firma, also the source of the French word ferme, meaning a fixed agreement, contract, from the classical Latin adjective firmus meaning strong, stout, firm. As in the medieval age virtually all manors were engaged in the business of agriculture, which was their principal revenue source, so to hold a manor by the tenure of "fee farm" became synonymous with the practice of agriculture itself.

History

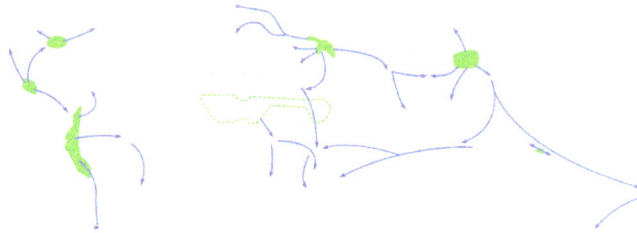

Map of the world showing approximate centers of origin of agriculture and its spread in prehistory: the Fertile Crescent (11,000 BP), the Yangtze and Yellow River basins (9,000 BP), and the New Guinea Highlands (9,000–6,000 BP), Central Mexico (5,000–4,000 BP), Northern South America (5,000–4,000 BP), sub-Saharan Africa (5,000–4,000 BP, exact location unknown), eastern North America (4,000–3,000 BP).

Farming has been innovated at multiple different points and places in human history. The transition from hunter-gatherer to settled, agricultural societies is called the Neolithic Revolution and first began around 12,000 years ago, near the beginning of the geological epoch of the Holocene around 12,000 years ago. It was the world's first historically verifiable revolution in agriculture. Subsequent step-changes in human farming practices were provoked by the British Agricultural Revolution in the 18th century, and the Green Revolution of the second half of the 20th century. Farming spread from the Middle East to Europe and by 4,000 BC people that lived in the central part of Europe were using oxen to pull plows and wagons.

Types of Farm

A farm may be owned and operated by a single individual, family, community, corporation or a company, may produce one or many types of produce, and can be a holding of any size from a fraction of a hectare to several thousand hectares.

A farm may operate under a monoculture system or with a variety of cereal or arable crops, which may be separate from or combined with raising livestock. Specialist farms are often denoted as such, thus a dairy farm, fish farm, poultry farm or mink farm.

Some farms may not use the word at all, hence vineyard (grapes), orchard (nuts and other fruit), market garden or "truck farm" (vegetables and flowers). Some farms may be denoted by their topographical location, such as a hill farm, while large estates growing cash crops such as cotton or coffee may be called plantations.

Many other terms are used to describe farms to denote their methods of production, as in collective, corporate, intensive, organic or vertical.

Other farms may primarily exist for research or education, such as an ant farm, and since farming is synonymous with mass production, the word "farm" may be used to describe wind power generation or puppy farm.

Specialized Farms

Dairy farm

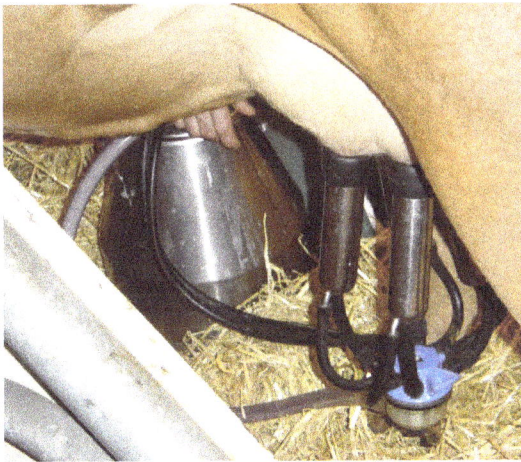

A milking machine in action

Dairy farming is a class of agriculture, where female cattle, goats, or other mammals are raised for their milk, which may be either processed on-site or transported to a dairy for processing and eventual retail sale. There are many breeds of cattle that can be milked some of the best producing ones include Holstein, Norwegian Red, Kostroma, Brown Swiss, and more.

In most Western countries, a centralized dairy facility processes milk and dairy products, such as cream, butter, and cheese. In the United States, these dairies are usually local companies, while in the southern hemisphere facilities may be run by very large nationwide or trans-national corporations (such as Fonterra).

Dairy farms generally sell male calves for veal meat, as dairy breeds are not normally satisfactory for commercial beef production. Many dairy farms also grow their own feed, typically including corn, alfalfa, and hay. This is fed directly to the cows, or stored as silage for use during the winter season. Additional dietary supplements are added to the feed to improve milk production.

Poultry Farm

Poultry farming

Poultry farms are devoted to raising chickens (egg layers or broilers), turkeys, ducks, and other fowl, generally for meat or eggs.

Pig Farm

A pig farm is one that specializes in raising pigs or hogs for bacon, ham and other pork products and may be free range, intensive, or both.

Prison Farm

Prison farms are farms which serve as prisons for people sentenced to hard labor by a court. On prison farms inmates run the important tasks of a farm and producing crops.

Ownership

Farm control and ownership has traditionally been a key indicator of status and power, especially in Medieval European agrarian societies. The distribution of farm ownership has historically been closely linked to form of government. Medieval feudalism was essentially a system that centralized control of farmland, control of farm labor and political power, while the early American democracy, in which land ownership was a prerequisite for voting rights, was built on relatively easy paths to individual farm ownership. However, the gradual modernization and mechanization of farming, which greatly increases both the efficiency and capital requirements of farming, has led to increasingly large farms. This has usually been accompanied by the decoupling of political power from farm ownership.

Forms of Ownership

In some societies (especially socialist and communist), collective farming is the norm, with either government ownership of the land or common ownership by a local group.

Especially in societies without widespread industrialized farming, tenant farming and sharecropping are common; farmers either pay landowners for the right to use farmland or give up a portion of the crops.

Farms Around the World

Americas

Farming near Klingerstown, Pennsylvania

A typical North American grain farm with farmstead in Ontario, Canada

The land and buildings of a farm are called the "farmstead". Enterprises where livestock are raised on rangeland are called ranches. Where livestock are raised in confinement on feed produced elsewhere, the term feedlot is usually used.

In 1910 there were 6,406,000 farms and 10,174,000 family workers; In 2000 there were only 2,172,000 farms and 2,062,300 family workers. The share of U.S. farms operated by women has risen steadily over recent decades, from 5 percent in 1978 to 14 percent by 2007.

In the United States, there are over three million migrant and seasonal farmworkers; 72% are foreign-born, 78% are male, they have an average age of 36 and average education of 8 years. Farmworkers make an average hourly rate of $9–10 per hour, compared to an average of over $18 per hour for nonfarm labor. Their average family income is under $20,000 and 23% live in families with incomes below the federal poverty level.

One-half of all farmworker families earn less than $10,000 per year, which is significantly below the 2005 U.S. poverty level of $19,874 for a family of four.

In 2007, corn acres are expected to increase by 15% because of the high demand for ethanol, both in and outside of the U.S. Producers are expecting to plant 90.5 million acres (366,000 km²) of corn, making it the largest corn crop since 1944.

Asia

Farmlands in Hebei province, China

Pakistan

According to the World Bank, "most empirical evidence indicates that land productivity on large farms in Pakistan is lower than that of small farms, holding other factors constant." Small farmers have "higher net returns per hectare" than large farms, according to farm household income data.

Nepal

Goat found in Nepal

Nepal is an agricultural country and about 80% of the total population are engaged in farming. Rice is mainly produced in Nepal along with fruits like apples. Dairy farming and poultry farming are also growing in Nepal.

Australia

Cows grazing on a farm in Victoria, Australia

Farming is a significant economic sector in Australia. A farm is an area of land used for primary production which will include buildings.

According to the UN, "green agriculture directs a greater share of total farming input expenditures towards the purchase of locally sourced inputs (e.g. labour and organic fertilisers) and a local multiplier effect is expected to kick in. Overall, green farming practices tend to require more labour inputs than conventional farming (e.g. from comparable levels to as much as 30 per cent more) (FAO 2007 and European Commission 2010), creating jobs in rural areas and a higher return on labour inputs."

Where most of the income is from some other employment, and the farm is really an expanded residence, the term hobby farm is common. This will allow sufficient size for recreational use but be very unlikely to produce sufficient income to be self-sustaining. Hobby farms are commonly around 2 hectares (4.9 acres) but may be much larger depending upon land prices (which vary regionally).

Often very small farms used for intensive primary production are referred to by the specialization they are being used for, such as a dairy rather than a dairy farm, a piggery, a market garden, etc. This also applies to feedlots, which are specifically developed to a single purpose and are often not able to be used for more general purpose (mixed) farming practices.

In remote areas farms can become quite large. As with estates in England, there is no defined size or method of operation at which a large farm becomes a station.

Europe

In the UK, farm as an agricultural unit, always denotes the area of pasture and other fields together with its farmhouse, farmyard and outbuildings. Large farms, or groups of farms under the same ownership, may be called an estate. Conversely, a small farm surrounding the owner's dwelling is called a smallholding and is generally focused on self-sufficiency with only the surplus being sold.

Traditional Dutch farmhouse

Farm Equipment

Farm equipment has evolved over the centuries from simple hand tools such as the hoe, through ox- or horse-drawn equipment such as the plough and harrow, to the modern highly-technical machinery such as the tractor, baler and combine harvester replacing what was a highly labour-intensive occupation before the Industrial revolution. Today much of the farm equipment used on both small and large farms is automated (e.g. using satellite guided farming).

As new types of high-tech farm equipment have become inaccessible to farmers that historically fixed their own equipment, Wired reports there is a growing backlash, due mostly to companies using intellectual property law to prevent farmers from having the legal right to fix their equipment (or gain access to the information to allow them to do it). This has encouraged groups such as Open Source Ecology and Farm Hack to begin to make open source hardware for agricultural machinery. In addition on a smaller scale Farmbot and the RepRap open source 3D printer community has begun to make open-source farm tools available of increasing levels of sophistication.

Herb Farm

An herb farm is usually a farm where herbs are grown for market sale. There is a case for the use of a small farm being dedicated to herb farming as the smaller farm is more efficient in terms of manpower usage and value of the crops on a per acre basis. In addition, the market for herbs is not as large as the more commercial crops, providing the justification for the small-scale herb farm. Herbs may be for culinary, medicinal or aromatic use, and sold fresh-cut or dried. Herbs may also be grown for their essential oils or as raw material for making herbal products. Many businesses calling themselves an herb farm sell potted herb plants for home gardens. Some herb farms also have gift shops, classes, and sometimes offer food for sale. In the United States, some herb farms belong to trade associations.

History

The rise of the herb farm stems from human interactions in agriculture, cultural preferences, growing conditions and availability of certain herbs. In early Egyptian times, herbs were grown for religious ceremonies, temple use, and in mummification, these herbs included: frankincense, myrrh, lotuses, poppies, cornflowers and chamomile. In Islamic cultures, herb gardens are also tied to buildings and are often enclosed behind walls and arranged in geometric patterns that maximize the use of space. Christian monasteries borrowed heavily from this same style and grew their herbs in a similar manner. As these monasteries expanded in influence and Christianity spread, the needs of the garden expanded as a means of creating a self-sufficient monastery. Culturally, these buildings are also associated with healing the sick and emphasis was placed on the growing of medicinal herbs. As cities grew, certain larger houses, and small plots of land were used to expand on the growing of herbs, both for cooking and medicinal properties. Universities began to explore the development of research into medicinal herbs, which required plots of land to be set aside to grow these herbs. In Greek gardens, the herb garden was paired with the bee garden, such that they became fused. The practice of keeping both types of herb farms together was carried over into European and American Colonial gardens.

In the colonial United States (in the time period from 1620-1840) herbs were grown for a variety of purposes, ranging from utilitarian, to vanity gardening. In Puritanical regions herbs were grown based on their functional use in cooking, medicines, and for a source of fragrance. Whereas the established towns the herb garden followed the trends of the time frame, sometimes they were present and other times not.

In rural China, herb growing was also important, and based on the cultural and scientific interests of that region.

Growing Process

Soil

Most herbs prefer a well drained, friable soil. This allows the roots to receive the water they need without the danger of rot and promotes a strong root system. There are a variety of soil amendments to be added if the soil in a specific area is not desirable for to promote good herb growth. For outdoor herb farming the soil must be prepared ahead of time by removing vegetation and analyzing soil for pH value and amount of fertilizer needed for proper growing conditions. In commercial industries, additional soil sterilization is used to eliminate common soil derived crop diseases, pests, and for the control of weeds. Additionally, crop rotation is important to diminish the possibility of common crop diseases.

Greenhouses/Nurseries

Greenhouse production allows for year-round growing of herbs, giving a level of control of temperature and water conditions inside of the greenhouse which is a desired outcome for the herb farmer. Other options for the greenhouse allow the grower to start seedlings early and then prepare them for transplant outdoors as the weather permits. The type of greenhouse used for growing depends on the long term needs of the herb farm, and consideration for utility costs should be factored when determining size and type of greenhouse installation necessary. Important factors for growing from the seedling stage to the full cycle of the plant are water, light, temperature control, mineral content of soil, control of the gases required in the process of photosynthesis, and pest and disease management.

Pest and Disease Control

Options for controlling pests of the herb farm are predator insects and insecticides. Some of the known predatory insects used on the herb farm are: ladybugs, aphid parasites, lacewings, mantids, and predatory mites. Insecticide use depends on the type of pest present on the herb farm. Common pests are aphids, whiteflies, and fungus gnats, each requiring a different type of pesticide and having different levels of difficulty to manage. Environmental management plans for pests include the use of a good insect screen, air movement, and ventilation: methods that reduce the stress on the plant. Common herb diseases are: botrytis, mildew, viruses, rusts, and root disease. Successful management of plant diseases include environmental controls similar to that of the pest management plan. Chemical means are determined based on the type of plant disease present in the herb.

Seeds and Cutting/Types of Propagation

Herb plants can be started from seed or purchased as a seedling. Common herbs grown from seed are basil, flat and curly leaved parsley, chives, dill, sage, thyme, rosemary, cilantro. Herbs can also be grown via vegetative means, rooting cuttings, division of the plant, bulbs, or tissue culture. The advantage of growing via a vegetative process gives you a plant that is exactly identical to the parent plant.

Farming Systems

Herb farms can range from in-house gardening hobby, mini-farming, and all the way to full-scale commercial operations. Start-up farms are generally smaller and focus on local markets, they can include organic, hydroponic and traditional growing systems. The small farm generally has one or two greenhouses, an area for storage, and an area to sell the product. Commercial operations focus on a larger scale of cash profit and may use some of the same growing systems, but can range vastly in their application of advanced technology and science. There are a variety of hydroponic systems for herb farming as herbs have a smaller root system and can be suited to the different types of hydroponic production. Growing herbs hydroponically is considered to be more ef-ficient, and to produce a higher quality product and can be seen in both the small farm and in commercial operations. In contrast, organic farming systems that additional make use of a greenhouse expand the growing season, is a fast growing niche market, and offers monetary value as a result. Organic certification criteria must be met based on soil content, fertilizer sources, and method of pest control.

Economics

The concept of starting up your own small, herb farms stems from the needs of lo-cal restaurants and at-home chefs demanding fresher, locally grown products. Home grown, small scale start ups often begin on a smaller piece of land and may also include areas of the house to be used to germinate seeds.

Herb garden at Kariwak Village in Tobago

Lavender growing on a farm

List of Culinary Herbs and Spices

Ajwain

Ajwain, ajowan Trachyspermum ammi, also known as Ajowan caraway, bishop's weed or carom, is an annual herb in the family Apiaceae. It originated in India and Pakistan. Both the leaves and the fruit (often mistakenly called seeds) of the plant are consumed by humans. The plant is also called bishop's weed, but this is a common name it shares with some other different plants. The "seed" (i.e., the fruit) is often confused with lovage "seed".

Description

Ajwain Fruit (Schizocarps)

The small fruits are pale brown schizocarps and have an oval shape, resembling caraway and cumin. It has a bitter and pungent taste, with a flavor similar to anise and oregano. They smell almost exactly like thyme because it also contains thymol, but is more aromatic and less subtle in taste, as well as slightly bitter and pungent. Even a small number of fruits tends to dominate the flavor of a dish.

Cultivation and Production

The plant is mainly cultivated in Iran and India. Rajasthan produced about 55% of India's total output in 2006.

Culinary uses

The fruits are rarely eaten raw; they are commonly dry-roasted or fried in ghee (clarified butter). This allows the spice to develop a more subtle and complex aroma. In Indian cuisine, it is often part of a chaunk, a mixture of spices fried in oil or butter, which is used to flavor lentil dishes. In Afghanistan, the fruits are sprinkled over bread and biscuits.

Medicinal uses

Ajwain is used as medicinal plant in traditional Ayurvedic medicine; primarily for stomach disorders such as indigestion, flatulence, and others but also for its supposed antispasmodic and carminative properties. In general the crushed fruits are applied externally as a poultice.

Essential Oil

Hydrodistillation of Ajwain fruits yields an essential oil consisting primarily of thymol, gamma-terpinene and p-cymene as well as more than 20 trace compounds (predominately terpenoids).

Bay Leaf

Indian bay leaf Cinnamomum tamala

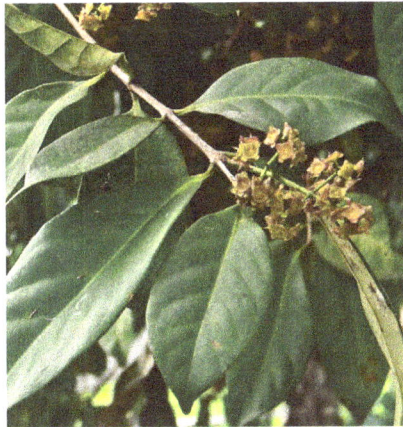

Indonesian bay leaf Syzygium polyanthum

Bay leaf (plural bay leaves) refers to the aromatic leaves of several plants used in cooking. These include:

- Bay laurel (Laurus nobilis, Lauraceae). Fresh or dried bay leaves are used in cooking for their distinctive flavor and fragrance. The leaves should be removed from the cooked food before eating. The leaves are often used to flavor soups, stews, braises and pâtés in Mediterranean cuisine. The fresh leaves are very mild and do not develop their full flavor until several weeks after picking and drying.

- California bay leaf – the leaf of the California bay tree (Umbellularia californica, Lauraceae), also known as California laurel, Oregon myrtle, and pepperwood, is similar to the Mediterranean bay laurel, but has a stronger flavor.

- Indian bay leaf or malabathrum (Cinnamomum tamala, Lauraceae) is somewhat similar in appearance to the leaves of bay laurel, but is culinarily quite different, having a fragrance and taste similar to cinnamon (cassia) bark, but milder.

- Indonesian bay leaf or Indonesian laurel (salam leaf, Syzygium polyanthum, Myrtaceae) is not commonly found outside of Indonesia; this herb is applied to meat and, less often vegetables.

- West Indian bay leaf, the leaf of the West Indian bay tree (Pimenta racemosa, Myrtaceae), used culinarily and to produce the cologne called bay rum.

- Mexican bay leaf (Litsea glaucescens, Lauraceae).

Chemical Constituents

The leaves contain about 1.3% essential oils (ol. lauri folii), consisting of 45% eucalyptol, 12% other terpenes, 8-12% terpinyl acetate, 3–4% sesquiterpenes, 3% methyleugenol, and other α- and β-pinenes, phellandrene, linalool, geraniol, and terpineol, contains lauric acid also.

Taste and Aroma

If eaten whole, bay leaves (Laurus nobilis) are pungent and have a sharp, bitter taste. As with many spices and flavorings, the fragrance of the bay leaf is more noticeable than its taste. When dried, the fragrance is herbal, slightly floral, and somewhat similar to oregano and thyme. Myrcene, which is a component of many essential oils used in perfumery, can be extracted from the bay leaf. They also contain the essential oil eugenol.

Uses

Bay leaves were used for flavoring by the ancient Greeks. They are a fixture in the cooking of many European cuisines (particularly those of the Mediterranean), as well as in the Americas. They are used in soups, stews, meat, seafood, vegetable dishes, and sauces. The leaves also flavor many classic French dishes. The leaves are most often used whole (sometimes in a bouquet garni) and removed before serving (they can be abrasive in the digestive tract). Thai cuisine employs bay leaf (Thai name bai kra wan) in a few Arab-influenced dishes, notably massaman curry.

In Indian and Pakistani cuisine, bay laurel leaves are sometimes used in place of Indian bay leaf, although they have a different flavor. They are most often used in rice dishes like biryani and as an ingredient in garam masala. Bay (laurel) leaves are frequently packaged as tejpatta (the Hindi term for Indian bay leaf), creating confusion between the two herbs.

In the Philippines, dried bay laurel leaves are added as a spice in the Filipino dish Adobo.

Bay leaves can also be crushed or ground before cooking. Crushed bay leaves impart more fragrance than whole leaves, but are more difficult to remove, and thus they are often used in a muslin bag or tea infuser. Ground bay laurel may be substituted for whole leaves, and does not need to be removed, but it is much stronger.

Bay leaves can also be used scattered in a pantry to repel meal moths, flies, cockroaches, mice, and silverfish.

Bay leaves have been used in entomology as the active ingredient in killing jars. The crushed, fresh, young leaves are put into the jar under a layer of paper. The vapors they release kill insects slowly but effectively, and keep the specimens relaxed and easy to mount. The leaves discourage the growth of molds. They are not effective for killing large beetles and similar specimens, but insects that have been killed in a cyanide killing jar can be transferred to a laurel jar to await mounting. It is not clear to what extent the effect is due to cyanide released by the crushed leaves, and to what extent other volatile products are responsible.

Safety

Some members of the laurel family, as well as the unrelated but visually similar mountain laurel and cherry laurel, have leaves that are poisonous to humans and livestock. While these plants are not sold anywhere for culinary use, their visual similarity to bay leaves has led to the oft-repeated belief that bay leaves should be removed from food after cooking because they are poisonous. This is not true – bay leaves may be eaten without toxic effect. However, they remain very stiff even after thorough cooking, and if swallowed whole or in large pieces, they may pose a risk of scratching the digestive tract or even causing choking. There are multiple cases of intestinal perforations caused by swallowing bay leaves, and they should not be swallowed or left in the food before serving to prevent the occurrence of a possibly fatal surgical emergency. Thus, most recipes that use bay leaves will recommend their removal after the cooking process has finished.

Coriander

Coriander also known as cilantro or Chinese parsley, is an annual herb in the family Apiaceae. All parts of the plant are edible, but the fresh leaves and the dried seeds are the parts most traditionally used in cooking.

Botanical Description

Coriander is native to regions spanning from southern Europe and northern Africa to southwestern Asia. It is a soft plant growing to 50 cm (20 in) tall. The leaves are variable in shape, broadly lobed at the base of the plant, and slender and feathery higher on the flowering stems. The flowers are borne in small umbels, white or very pale pink, asymmetrical, with the petals pointing away from the center of the umbel longer (5–6 mm or 0.20–0.24 in) than those pointing toward it (only 1–3 mm or 0.039–0.118 in long). The fruit is a globular, dry schizocarp 3–5 mm (0.12–0.20 in) in diameter.

History

Coriander plants

Coriander grows wild over a wide area of Western Asia and southern Europe, prompting the comment, "It is hard to define exactly where this plant is wild and where it only recently established itself." Fifteen desiccated mericarps were found in the Pre-Pottery Neolithic B level of the Nahal Hemar Cave in Israel, which may be the oldest archaeological find of coriander. About half a litre (a pint) of coriander mericarps was recovered from the tomb of Tutankhamen, and because this plant does not grow wild in Egypt, Zohary and Hopf interpret this find as proof that coriander was cultivated by the ancient Egyptians.

Coriander seems to have been cultivated in Greece since at least the second millennium BC. One of the Linear B tablets recovered from Pylos refers to the species as being cultivated for the manufacture of perfumes, it apparently was used in two forms: as a spice for its seeds and as a herb for the flavour of its leaves. This appears to be confirmed by archaeological evidence from the same period; the large quantities of the species retrieved from an Early Bronze Age layer at Sitagroi in Macedonia could point to cultivation of the species at that time.

Coriander was brought to the British colonies in North America in 1670, and was one of the first spices cultivated by early settlers.

Uses

All parts of the plant are edible, but the fresh leaves and the dried seeds are the parts most traditionally used in cooking. Coriander is used in cuisines throughout the world.

Leaves

Coriander leaves

The leaves are variously referred to as coriander leaves, fresh coriander, dhania, Chinese parsley, or (in the US and commercially in Canada) cilantro.

Coriander potentially may be confused with culantro (Eryngium foetidum L.), an apiacea like coriander (Coriandrum sativum L.), but from a different genus. Culantro has a distinctly different spiny appearance, a more potent volatile leaf oil and a stronger aroma.

The leaves have a different taste from the seeds, with citrus overtones. However, some people find the leaves to have an unpleasant soapy taste or a rank smell and avoid them.

The fresh leaves are an ingredient in many South Asian foods (such as chutneys and salads); in Chinese and Thai dishes; in Mexican cooking, particularly in salsa and guacamole and as a garnish; and in salads in Russia and other CIS countries. Chopped coriander leaves are a garnish on Indian dishes such as dal. As heat diminishes their flavour, coriander leaves are often used raw or added to the dish immediately before serving. In Indian and Central Asian recipes, coriander leaves are used in large amounts and cooked until the flavour diminishes. The leaves spoil quickly when removed from the plant, and lose their aroma when dried or frozen.

Fruits

The dry fruits are known as coriander seeds. The word "coriander" in food preparation may refer solely to these seeds (as a spice), rather than to the plant. The seeds have a lemony citrus flavour when crushed, due to terpenes linalool and pinene. It is described as warm, nutty, spicy, and orange-flavoured.

Dried coriander fruits, often called "coriander seeds" when used as a spice

Coriander roots

The variety C. s. vulgare has a fruit diameter of 3–5 mm (0.12–0.20 in), while var. C. s. microcarpum fruits have a diameter of 1.5–3 mm (0.06–0.12 in). Large-fruited types are grown mainly by tropical and subtropical countries, e.g. Morocco, India, and Australia, and contain a low volatile oil content (0.1-0.4%). They are used extensively for grinding and blending purposes in the spice trade. Types with smaller fruit are produced in temperate regions and usually have a volatile oil content around 0.4-1.8%, so are highly valued as a raw material for the preparation of essential oil.

Food Applications

Coriander is commonly found both as whole dried seeds and in ground form. Roasting or heating the seeds in a dry pan heightens the flavour, aroma, and pungency. Ground coriander seed loses flavour quickly in storage and is best ground fresh. Coriander seed is a spice in garam masala and Indian curries which often employ the ground fruits in generous amounts together with cumin, acting as a thickener in a mixture called dhana jeera.

Roasted coriander seeds, called dhana dal, are eaten as a snack. They are the main ingredient of the two south Indian dishes: sambhar and rasam.

Outside of Asia, coriander seed is used widely in the process for pickling vegetables. In Germany and South Africa, the seeds are used while making sausages. In Russia and

Central Europe, coriander seed is an occasional ingredient in rye bread (e.g. Borodinsky bread), as an alternative to caraway.

The Zuni people of North America have adapted it into their cuisine, mixing the powdered seeds ground with chile and using it as a condiment with meat, and eating leaves as a salad.

Coriander seeds are used in brewing certain styles of beer, particularly some Belgian wheat beers#Witbier. The coriander seeds are used with orange peel to add a citrus character.

Research

One preliminary study showed coriander essential oil to inhibit Gram-positive and Gram-negative bacteria, including Staphylococcus aureus, Enterococcus faecalis, Pseudomonas aeruginosa, and Escherichia coli.

Roots

Having a deeper, more intense flavor than the leaves, coriander roots are used in a variety of Asian cuisines, especially in Thai dishes such as soups or curry pastes.

Flowering Coriander for Aphid Control

In the Salinas Valley of California, aphids have been one of the worst pest in the lettuce fields. The USDA Cooperative Extension Service has been investigating organic methods for aphid control, and experimented with coriander plants and Alyssum plants; when intercropped with the lettuce and allowed to flower, they attract beneficial insects such as hoverflies, the larvae of which eat up to 150 aphids per day before they mature into flying adults.

Nutrients

The nutritional profile of coriander seeds is different from the fresh stems or leaves. Leaves are particularly rich in vitamin A, vitamin C and vitamin K, with moderate content of dietary minerals (table above). Although seeds generally have lower content of vitamins, they do provide significant amounts of dietary fiber, calcium, selenium, iron, magnesium and manganese.

Taste and Smell

Different people may perceive the taste of coriander leaves differently. Those who enjoy it say it has a refreshing, lemony or lime-like flavor, while those who dislike it have a strong aversion to its taste and smell, characterizing it as soapy or rotten.

Studies also show variations in preference among different ethnic groups: 21% of East Asians, 17% of Caucasians, and 14% of people of African descent expressed a dislike for coriander, but among the groups where coriander is popular in their cuisine, only 7% of South Asians, 4% of Hispanics, and 3% of Middle Eastern subjects expressed a dislike.

Flowers of Coriandrum sativum

Twin studies have shown that 80% of identical twins shared the same preference for the herb, but fraternal twins agreed only about half the time, strongly suggesting a genetic component to the preference. In a genetic survey of nearly 30,000 people, two genetic variants linked to perception of coriander have been found, the most common of which is a gene involved in sensing smells. The gene, OR6A2, lies within a cluster of olfactory-receptor genes, and encodes a receptor that is highly sensitive to aldehyde chemicals. Flavor chemists have found that the coriander aroma is created by a half-dozen or so substances, and most of these are aldehydes. Those who dislike the taste are sensitive to the offending unsaturated aldehydes, while simultaneously may also be unable to detect the aromatic chemicals that others find pleasant. Association between its taste and several other genes, including a bitter-taste receptor, have also been found.

Similar Plants

Other herbs are used where they grow in much the same way as coriander leaves.

- Eryngium foetidum has a similar, but more intense, taste. Known as culantro, it is found in Mexico, South America, and the Caribbean.

- Persicaria odorata is commonly called Vietnamese coriander, or rau răm. The leaves have a similar odour and flavour to coriander. It is a member of the Polygonaceae, or buckwheat family.

- Papaloquelite is one common name for Porophyllum ruderale subsp. macrocephalum, a member of the Compositae or Asteraceae, the sunflower family. This species is found growing wild from Texas to Argentina.

Allergy

Coriander can produce an allergic reaction in some people.

Fennel

Fennel (Foeniculum vulgare) is a flowering plant species in the carrot family. It is a hardy, perennial herb with yellow flowers and feathery leaves. It is indigenous to the shores of the Mediterranean but has become widely naturalized in many parts of the world, especially on dry soils near the sea-coast and on riverbanks.

It is a highly aromatic and flavorful herb with culinary and medicinal uses and, along with the similar-tasting anise, is one of the primary ingredients of absinthe. Florence fennel or finocchio is a selection with a swollen, bulb-like stem base that is used as a vegetable.

Fennel is used as a food plant by the larvae of some Lepidoptera species including the mouse moth and the anise swallowtail.

Etymology and Names

The word "fennel" developed from the Middle English fenel or fenyl. This came from the Old English fenol or finol, which in turn came from the Latin feniculum or foeniculum, the diminutive of fenum or faenum, meaning "hay". The Latin word for the plant was ferula, which is now used as the genus name of a related plant.

Cultural References

As Old English finule, fennel is one of the nine plants invoked in the pagan Anglo-Saxon Nine Herbs Charm, recorded in the 10th century.

The Greek name for fennel is marathon or marathos, and the place of the famous battle of Marathon (whence marathon, the subsequent sports event), literally means a plain with fennels. The word is first attested in Mycenaean Linear B form as ma-ra-tu-wo.

The name Funchal was given to their new town by the first settlers on Madeira due to the abundance of wild fennel, from the Portuguese word funcho (fennel) and the suffix -al.

Fennel, from Koehler's Medicinal-plants (1887)

Longfellow's 1842 poem "The Goblet of Life" repeatedly refers to the plant and mentions its purported ability to strengthen eyesight:

> Above the lower plants it towers,
>
> The Fennel with its yellow flowers;
>
> And in an earlier age than ours
>
> Was gifted with the wondrous powers
>
> Lost vision to restore.

Appearance

Fennel, Foeniculum vulgare, is a perennial herb. It is erect, glaucous green, and grows to heights of up to 2.5 m, with hollow stems. The leaves grow up to 40 cm long; they are finely dissected, with the ultimate segments filiform (threadlike), about 0.5 mm wide. (Its leaves are similar to those of dill, but thinner.) The flowers are produced in terminal compound umbels 5–15 cm wide, each umbel section having 20–50 tiny yellow flowers on short pedicels. The fruit is a dry seed from 4–10 mm long, half as wide or less, and grooved.

Fennel flowerheads

Fennel seeds

Cultivation and Uses

Fennel is widely cultivated, both in its native range and elsewhere, for its edible, strongly flavored leaves and fruits. Its aniseed flavor comes from anethole, an aromatic compound also found in anise and star anise, and its taste and aroma are similar to theirs, though usually not as strong.

Florence fennel (Foeniculum vulgare Azoricum Group; syn. F. vulgare var. azoricum) is a cultivar group with inflated leaf bases which form a bulb-like structure. It is of cultivated origin, and has a mild anise-like flavor, but is sweeter and more aromatic. Florence fennel plants are smaller than the wild type. The inflated leaf bases are eaten as a vegetable, both raw and cooked. Several cultivars of Florence fennel are also known by several other names, notably the Italian name finocchio. In North American supermarkets, it is often mislabeled as "anise".

Foeniculum vulgare 'Purpureum' or 'Nigra', "bronze-leaved" fennel, is widely available as a decorative garden plant.

Fennel has become naturalized along roadsides, in pastures, and in other open sites in many regions, including northern Europe, the United States, southern Canada, and much of Asia and Australia. It propagates well by seed, and is considered an invasive species and a weed in Australia and the United States. In western North America, fennel can be found from the coastal and inland wildland-urban interface east into hill and mountain areas, excluding desert habitats.

Florence fennel bulbs

Florence fennel is one of the three main herbs used in the preparation of absinthe, an alcoholic mixture which originated as a medicinal elixir in Switzerland and became, by the late 19th century, a popular alcoholic drink in France and other countries.

Nutrition

In a 100 gram amount, fennel seeds provide 345 calories and are a rich source (more than 19% of the Daily Value, DV) of protein, dietary fiber, B vitamins and several dietary minerals, especially calcium, iron, magnesium and manganese, all of which exceed 100% DV (table). Fennel seeds are 52% carbohydrates, 15% fat, 40% dietary fiber, 16% protein and 9% water (table).

Uses

Sugar-coated and uncoated fennel seeds are used in India and Pakistan in mukhwas, an after-meal snack and breath freshener.

Cuisine

The bulb, foliage, and seeds of the fennel plant are used in many of the culinary traditions of the world. The small flowers of wild fennel (known as fennel "pollen") are the most potent form of fennel, but also the most expensive. Dried fennel seed is an aromatic, anise-flavored spice, brown or green in color when fresh, slowly turning a dull grey as the seed ages. For cooking, green seeds are optimal. The leaves are delicately flavored and similar in shape to those of dill. The bulb is a crisp vegetable that can be sautéed, stewed, braised, grilled, or eaten raw. Young tender leaves are used for garnishes, as a salad, to add flavor to salads, to flavor sauces to be served with puddings, and also in soups and fish sauce.

Fennel seeds are sometimes confused with those of anise, which are similar in taste and appearance, though smaller. Fennel is also used as a flavoring in some natural toothpastes. The seeds are used in cookery and sweet desserts.

Many cultures in India and neighboring countries (where it is known as saunf in Hindi), Afghanistan, Iran, and the Middle East use fennel seed in cooking as one of the most important spices in Kashmiri Pandit and Gujarati cooking. It is an essential ingredient of the Assamese/Bengali/Oriya spice mixture panch phoron and in Chinese five-spice powders. In many parts of India and Pakistan, roasted fennel seeds are consumed as mukhwas, an after-meal digestive and breath freshener, or candied as comfit.

Fennel leaves are used in some parts of India as leafy green vegetables either by themselves or mixed with other vegetables, cooked to be served and consumed as part of a meal. In Syria and Lebanon, the young leaves are used to make a special kind of egg omelette (along with onions and flour) called ijjeh.

Many egg, fish, and other dishes employ fresh or dried fennel leaves. Florence fennel is a key ingredient in some Italian and German salads, often tossed with chicory and avocado, or it can be braised and served as a warm side dish. It may be blanched or marinated, or cooked in risotto.

Fennel seeds are the primary flavor component in Italian sausage. In Spain, the stems of the fennel plant are used in the preparation of pickled eggplants, berenjenas de Almagro. An herbal tea or tisane can be made from fennel.

On account of its aromatic properties, fennel fruit forms one of the ingredients of the well-known compound liquorice powder. In the Indian subcontinent, fennel seeds are also eaten raw, sometimes with a sweetener.

In Israel, fennel salad is made of chopped fennel bulbs flavored with salt, black pepper, lemon juice, parsley, olive oil and sometimes sumac.

Chemistry

Foeniculoside I is a stilbenoid. It is a glucoside of the stilbene trimer cis-miyabenol C. It can be found in Foeniculi fructus (fruit of F. vulgare).

Production

Fennel flower from Naduvil, India

India is the leader in production of anise, badian (star anise), fennel and coriander.

Top ten anise, badian, fennel & coriander producers — 11 June 2008		
Country	Production (tonnes per year)	Footnote
India	110,000	F
Mexico	49,688	F
China	40,000	F
Iran	30,000	F
Bulgaria	28,100	F
Syria	27,700	
Morocco	23,000	F
Egypt	22,000	F
Canada	11,000	F
Afghanistan	10,000	F
World	415,027	A
No symbol = official figure, P = official figure, F = FAO estimate, * = Unofficial/Semi-official/mirror data, C = Calculated figure A = Aggregate (may include official, semi-official or estimates);		
Source: Food And Agricultural Organization of United Nations: Economic And Social Department: The Statistical Division		

Similar Species

Many species in the family Apiaceae or Umbelliferae are superficially similar to fennel, and some, such as poison hemlock, are toxic, so it is unwise, and potentially extremely dangerous, to use any part of any of these plants as an herb or vegetable unless it can be positively identified as being edible.

Dill, coriander, and caraway are similar-looking herbs, but shorter-growing than fennel, reaching only 40–60 cm (16–24 in). Dill has thread-like, feathery leaves and yellow flowers; coriander and caraway have white flowers and finely divided leaves (though not as fine as dill or fennel) and are also shorter-lived (being annual or biennial plants). The superficial similarity in appearance between these may have led to a sharing of names and etymology, as in the case of meridian fennel, a term for caraway.

Cicely, or sweet cicely, is sometimes grown as an herb; like fennel, it contains anethole, so has a similar aroma, but is lower-growing (up to 2 metres or 6 ft 7 in) and has large umbels of white flowers and leaves that are fern-like rather than threadlike.

Giant fennel (Ferula communis) is a large, coarse plant, with a pungent aroma, which grows wild in the Mediterranean region and is only occasionally grown in gardens elsewhere. Other species of the genus Ferula are also commonly called giant fennel, but they are not culinary herbs.

In North America, fennel may be found growing in the same habitat and alongside natives osha (Ligusticum porteri) and Lomatium species, useful medicinal relatives in the parsley family.

Most Lomatium species have yellow flowers like fennel, but some are white flowered and resemble poison hemlock. Lomatium is an important historical food plant of Native Americans known as 'biscuit root'. Most Lomatium spp. have finely divided, hair-like leaves; their roots have a delicate rice-like odor, unlike the musty odor of hemlock. Lomatium species tend to prefer dry rocky soils devoid of organic material.

Square Foot Gardening

Square Foot Garden in Raised Bed

Square foot gardening is the practice of dividing the growing area into small square sections (typically 12" on a side, hence the name). The aim is to assist the planning and creating of a small but intensively planted vegetable garden. It results in a simple and orderly gardening system, from which it draws much of its appeal. Mel Bartholomew coined the term "square foot gardening" in his 1981 book of the same name.

Overview

The phrase "square foot gardening" was popularized by Mel Bartholomew in a 1981 Rodale Press book and subsequent PBS television series. Bartholomew used a 12' by 12' square with a grid that divided it into 9 squares with equal lengths of 4 feet on each side. Each of these 4' by 4' squares was then invisibly divided into sixteen one foot squares that were each planted with a different species. In smaller square gardens the grids may simply serve as a way to divide the garden but in larger gardens the grids can be made wide enough to be used as narrow walkways. Bartholomew recommends careful spacing of seeds rather than planting the entire seed packet so that fewer but stronger plants will grow.

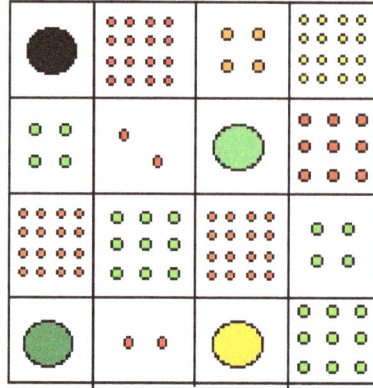

A basic, 4x4, 16-unit "square-foot garden."

To encourage a variety of different crops over time, each square would be used for a different kind of plant, the number of plants per square depending on an individual plant's size. For example, a single tomato plant might take a full square, as might herbs such as oregano, basil or mint, while most strawberry plants could be planted four per square, and up to sixteen per square of plants such as radish. Tall or climbing plants such as maize or pole beans might be planted in a northern row (south in the southern hemisphere) so as not to shade other plants, and supported with lattice or netting.

One advantage of densely planted crops is that they can form a living mulch, and also prevent weeds from establishing or even germinating. Also, natural insect repellent methods such as companion planting (i.e. planting marigolds or other naturally pest-repelling plants) become more efficient in a close space, which may reduce the need to use pesticides. The large variety of crops in a small space also prevents plant diseases from spreading easily

Since the beds are typically small, making covers or cages to protect plants from pests, cold, or sun is more practical than with larger gardens. To extend the growing season of a square foot garden, a cold/hot frame may be built around the SFG, and by facing the cold/hot frame south, the SFG captures more light and heat during the colder months of spring and winter.

In 2006, Bartholemew updated the concept with the book "All New Square Foot Gardening", which advocates growing in raised beds instead of the ground. The rationale is that by using potting soil instead of dirt, one will enjoy the benefits of having perfect soil conditions from day one, instead of needing to undertake a period of soil improvement.

Forest Gardening

Forest gardening is a low-maintenance sustainable plant-based food production and agroforestry system based on woodland ecosystems, incorporating fruit and nut trees, shrubs, herbs, vines and perennial vegetables which have yields directly useful to humans. Making use of companion planting, these can be intermixed to grow in a succession of layers, to build a woodland habitat.

Robert Hart's forest garden in Shropshire

Forest gardening is a prehistoric method of securing food in tropical areas. In the 1980s, Robert Hart coined the term "forest gardening" after adapting the principles and applying them to temperate climates.

History

Forest gardens are probably the world's oldest form of land use and most resilient agro-ecosystem. They originated in prehistoric times along jungle-clad river banks and in the wet foothills of monsoon regions. In the gradual process of families improving their immediate environment, useful tree and vine species were identified, protected and improved whilst undesirable species were eliminated. Eventually superior foreign species were selected and incorporated into the gardens.

Forest gardens are still common in the tropics and known by various names such as: home gardens in Kerala in South India, Nepal, Zambia, Zimbabwe and Tanzania; Kandyan forest gardens in Sri Lanka; huertos familiares, the "family orchards" of Mexico; and pekarangan, the gardens of "complete design", in Java. These are also called agro-forests and, where the wood components are short-statured, the term shrub garden is employed. Forest gardens have been shown to be a significant source of income and food security for local populations.

Robert Hart adapted forest gardening for the United Kingdom's temperate climate during the 1980s. His theories were later developed by Martin Crawford from the Agroforestry Research Trust and various permaculturalists such as Graham Bell, Patrick Whitefield, Dave Jacke and Geoff Lawton.

In Tropical Climates

Forest gardens, or home gardens, are common in the tropics, using intercropping to cultivate trees, crops, and livestock on the same land. In Kerala in south India as well as in northeastern India, the home garden is the most common form of land use and is also found in Indonesia. One example combines coconut, black pepper, cocoa and pineapple. These gardens exemplify polyculture, and conserve much crop genetic diversity and heirloom plants that are not found in monocultures. Forest gardens have been loosely compared to the religious concept of the Garden of Eden.

Americas

The BBC's Unnatural Histories claimed that the Amazon rainforest, rather than being a pristine wilderness, has been shaped by humans for at least 11,000 years through practices such as forest gardening and terra preta. This was also explored in the bestselling book 1491 by author Charles C. Mann. Since the 1970s, numerous geoglyphs have also been discovered on deforested land in the Amazon rainforest, furthering the evidence about Pre-Columbian civilizations.

On the Yucatán Peninsula, much of the Maya food supply was grown in "orchard-gardens", known as pet kot. The system takes its name from the low wall of stones (pet meaning circular and kot wall of loose stones) that characteristically surrounds the gardens.

Africa

In many African countries, for example Zambia, Zimbabwe, Ethiopia and Tanzania, gardens are widespread in rural, periurban and urban areas and they play an essential role in establishing food security. Most well known are the Chaga or Chagga gardens on the slopes of Mt. Kilimanjaro in Tanzania. These are an excellent example of an agroforestry system. In many countries, women are the main actors in home gardening and food is mainly produced for subsistence. In North-Africa, oasis layered gardening with palm trees, fruit trees and vegetables is a traditional type of forest garden.

Nepal

In Nepal, the Ghar Bagaincha, literally "home garden", refers to the traditional land use system around a homestead, where several species of plants are grown and maintained by household members and their products are primarily intended for the family consumption (Shrestha et al., 2002). The term "home garden" is often considered synonymous to the kitchen garden. However, they differ in terms of function, size, diversity, composition and features (Sthapit et al., 2006). In Nepal, 72% of households have home gardens of an area 2–11% of the total land holdings (Gautam et al., 2004). Because of their small size, the government has never identified home gardens as an important unit of food production and they thereby remain neglected from research and development. However, at the household level the system is very important as it is an important source of quality food and nutrition for the rural poor and, therefore, are important contributors to the household food security and livelihoods of farming communities in Nepal. The gardens are typically cultivated with a mixture of annual and perennial plants that can be harvested on a daily or seasonal basis. Biodiversity that has an immediate value is maintained in home gardens as women and children have easy access to preferred food. Home gardens, with their intensive and multiple uses, provide a safety net for households when food is scarce. These gardens are not only important sources of food, fodder, fuel, medicines, spices, herbs, flowers, construction materials and income in many countries, they are also important for the in situ conservation of a wide range of unique genetic resources for food and agriculture (Subedi et al., 2004). Many uncultivated, as well as neglected and underutilised species could make an important contribution to the dietary diversity of local communities (Gautam et al., 2004).

In addition to supplementing diet in times of difficulty, home gardens promote whole-family and whole-community involvement in the process of providing food. Children, the elderly, and those caring for them can participate in this infield agriculture, incorporating it with other household tasks and scheduling. This tradition has existed in many cultures around the world for thousands of years.

In Mediterranean Climates

The Mediterranean climate has long, hot, rainless summers and relatively short, cool, rainy winters (Köppen climate classification Csa). Its climate conditions are highly variable within an area and modified locally by altitude, latitude, and the proximity to the Mediterranean. In the 1950s the Forest Research Department of the Ministry of Agriculture founded a botanical forest garden in the Sharon region in Israel, the Ilanot Forest. As the only one of its kind in Israel, it harbours more than 750 species of trees from locations all over the world, including the Japanese sago palm cycas revoluta, fig trees (ficus glomerata), stone pine trees (pinus pinea) that produce tasty pine nuts and adds to the biodiversity of Israel.

In Temperate Climates

Robert Hart coined the term "forest gardening" during the 1980s. Hart began farming at Wenlock Edge in Shropshire with the intention of providing a healthy and therapeutic environment for himself and his brother Lacon. Starting as relatively conventional smallholders, Hart soon discovered that maintaining large annual vegetable beds, rearing livestock and taking care of an orchard were tasks beyond their strength. However, a small bed of perennial vegetables and herbs he planted was looking after itself with little intervention.

Robert Hart, forest gardening pioneer

Following Hart's adoption of a raw vegan diet for health and personal reasons, he replaced his farm animals with plants. The three main products from a forest garden are fruit, nuts and green leafy vegetables. He created a model forest garden from a 0.12 acre (500 m²) orchard on his farm and intended naming his gardening method ecological horticulture or ecocultivation. Hart later dropped these terms once he became aware that agroforestry and forest gardens were already being used to describe similar systems in other parts of the world. He was inspired by the forest farming methods of Toyohiko Kagawa and James Sholto Douglas, and the productivity of the Keralan home gardens as Hart explains:

From the agroforestry point of view, perhaps the world's most advanced country is the Indian state of Kerala, which boasts no fewer than three and a half million forest gardens...As an example of the extraordinary intensivity of cultivation of some forest gardens, one plot of only 0.12 hectares (0.30 acres) was found by a study group to have twenty-three young coconut palms, twelve cloves, fifty-six bananas, and forty-nine pineapples, with thirty pepper vines trained up its trees. In addition, the small holder grew fodder for his house-cow.

Seven-layer System

1. CANOPY (LARGE FRUIT & NUT TREES)
2. LOW TREE LAYER (DWARF FRUIT TREES)
3. SHRUB LAYER (CURRANTS & BERRIES)
4. HERBACEOUS (COMFREYS, BEETS, HERBS)
5. RHIZOSPHERE (ROOT VEGETABLES)
6. SOIL SURFACE (GROUND COVER, EG, STRAWBERRY, ETC)
7. VERTICAL LAYER (CLIMBERS, VINES)

THE FOREST GARDEN: A SEVEN LEVEL BENEFICIAL GUILD

The seven layers of the forest garden

Robert Hart pioneered a system based on the observation that the natural forest can be divided into distinct levels. He used intercropping to develop an existing small orchard of apples and pears into an edible polyculture landscape consisting of the following layers:

1. 'Canopy layer' consisting of the original mature fruit trees.

2. 'Low-tree layer' of smaller nut and fruit trees on dwarfing root stocks.

3. 'Shrub layer' of fruit bushes such as currants and berries.

4. 'Herbaceous layer' of perennial vegetables and herbs.

5. 'Rhizosphere' or 'underground' dimension of plants grown for their roots and tubers.

6. 'Ground cover layer' of edible plants that spread horizontally.

7. 'Vertical layer' of vines and climbers.

A key component of the seven-layer system was the plants he selected. Most of the traditional vegetable crops grown today, such as carrots, are sun loving plants not well selected for the more shady forest garden system. Hart favoured shade tolerant perennial vegetables.

Further Development

The Agroforestry Research Trust (ART), managed by Martin Crawford, runs experimental forest gardening projects on a number of plots in Devon, United Kingdom. Crawford describes a forest garden as a low-maintenance way of sustainably producing food and other household products.

Ken Fern had the idea that for a successful temperate forest garden a wider range of edible shade tolerant plants would need to be used. To this end, Fern created the organisation Plants for a Future (PFAF) which compiled a plant database suitable for such a system. Fern used the term woodland gardening, rather than forest gardening, in his book Plants for a Future.

The Movement for Compassionate Living (MCL) promote forest gardening and other types of vegan organic gardening to meet society's needs for food and natural resources. Kathleen Jannaway, the founder of MCL, wrote a book outlining a sustainable vegan future called Abundant Living in the Coming Age of the Tree in 1991. In 2009, the MCL provided a grant of £1,000 to the Bangor Forest Garden project in Gwynedd, North West Wales.

Kevin Bradley coined the phrase "Edible Forest" in the 1980s as the name of his nursery, garden, and orchard on 5 acres in the frigid zone 3 pine forests of northern Wisconsin. Among 3 options, he chose "Edible Forest" because it "evokes at once an ethereal, spiritual, and magical image", of Disney- like "Forest of No Return"; of the biblical "Garden of Eden". This image was perfectly in line with his ongoing experiment begun in 1985 in what he calls a closed loop human environment, combining multi- story tree and field crop "garden/orchards" for maximum beauty and use of space, someday to be very useful in an ever shrinking world. "The name, at the same time, with its irrational first impression (of course we can't eat a forest), forces the mind to think, if just a little bit, about its inference and thus sticks in our memories". It appeared from Bradley's research that the two words had, prior to the 80's, never been put together before as a noun phrase but which by today, after more than two decades of Bradley's "Edible Forest Nursery" and the 2005 text by Jacke and Toensmeirer's- "Edible Forest Gardens", has grown into a movement and little "Edible Forests" all over the world.

In 2005, Dave Jacke and Eric Toensmeier's two-volume Edible Forest Gardens provided a deeply researched reference focused on North American forest gardening climates, habitats, and species. The book attempts to ground forest gardening deeply in ecological science. The Apios Institute wiki grew out of their work, and seeks to document and share the experience of people around the world working with the species in polycultures.

Permaculture

Bill Mollison, who coined the term permaculture, visited Robert Hart at his forest garden in Wenlock Edge in October 1990. Hart's seven-layer system has since been adopted as a common permaculture design element.

Numerous permaculturalists are proponents of forest gardens, or food forests, such as Graham Bell, Patrick Whitefield, Dave Jacke, Eric Toensmeier and Geoff Lawton. Bell started building his forest garden in 1991 and wrote the book The Permaculture Garden in 1995, Whitefield wrote the book How to Make a Forest Garden in 2002, Jacke and Toensmeier co-authored the two volume book set Edible Forest Gardening in 2005, and Lawton presented the film Establishing a Food Forest in 2008.

Austrian Sepp Holzer practices "Holzer Permaculture" on his Krameterhof farm, at varying altitudes ranging from 1,100 to 1,500 metres above sea level. His designs create micro-climates with rocks, ponds and living wind barriers, enabling the cultivation of a variety of fruit trees, vegetables and flowers in a region that averages 4 °C, and with temperatures as low as -20 °C in the winter.

Projects

El Pilar on the Belize-Guatemala border features a forest garden to demonstrate traditional Maya agricultural practices. A further 1-acre model forest garden, called Känan K'aax (meaning well-tended garden in Mayan), is being funded by the National Geographic Society and developed at Santa Familia Primary School in Cayo.

In the United States the largest known food forest on public land is believed to be the 7-acre Beacon Food Forest in Seattle, Washington. Other forest garden projects include those at the Central Rocky Mountain Permaculture Institute in Basalt, Colorado and Montview Neighborhood farm in Northampton, Massachusetts.

In Canada food forester Richard Walker has been developing and maintaining food forests in the province of British Columbia for over 30 years. He developed a 3-acre food forest that when at maturity provided raw materials for a nursery and herbalism business as well as food for his family. The Living Centre have developed various forest garden projects in Ontario.

In the United Kingdom, other than those run by the Agroforestry Research Trust (ART), there are numerous forest garden projects such as the Bangor Forest Garden in Gwynedd, North West Wales. Martin Crawford from ART administers the Forest Garden Network, an informal network of people and organisations around the world who are cultivating their own forest gardens.

Organic Horticulture

An organic garden on a school campus

Organic horticulture is the science and art of growing fruits, vegetables, flowers, or ornamental plants by following the essential principles of organic agriculture in soil building and conservation, pest management, and heirloom variety preservation.

The Latin words hortus (garden plant) and cultura (culture) together form horticulture, classically defined as the culture or growing of garden plants. Horticulture is also sometimes defined simply as "agriculture minus the plough." Instead of the plough, horticulture makes use of human labour and gardener's hand tools, although some small machine tools like rotary tillers are commonly employed now.

General

Mulches, cover crops, compost, manures, vermicompost, and mineral supplements are soil-building mainstays that distinguish this type of farming from its commercial coun-terpart. Through attention to good healthy soil condition, it is expected that insect, fungal, or other problems that sometimes plague plants can be minimized. However, pheromone traps, insecticidal soap sprays, and other pest-control methods available to organic farmers are also utilized by organic horticulturists.

Horticulture involves five areas of study. These areas are floriculture (includes pro-duction and marketing of floral crops), landscape horticulture (includes production, marketing and maintenance of landscape plants), olericulture (includes production and marketing of vegetables), pomology (includes production and marketing of fruits), and postharvest physiology (involves maintaining quality and preventing spoilage of horticultural crops). All of these can be, and sometimes are, pursued according to the principles of organic cultivation.

Organic horticulture (or organic gardening) is based on knowledge and techniques gathered over thousands of years. In general terms, organic horticulture involves nat-ural processes, often taking place over extended periods of time, and a sustainable, holistic approach - while chemical-based horticulture focuses on immediate, isolated effects and reductionist strategies.

Organic Gardening Systems

There are a number of formal organic gardening and farming systems that prescribe specific techniques. They tend to be more specific than, and fit within, general organic standards. Forest gardening, a fully organic food production system which dates from prehistoric times, is thought to be the world's oldest and most resilient agroecosystem.

Biodynamic farming is an approach based on the esoteric teachings of Rudolf Stein-er. The Japanese farmer and writer Masanobu Fukuoka invented a no-till system for small-scale grain production that he called Natural Farming. French intensive garden-ing and biointensive methods and SPIN Farming (Small Plot INtensive) are all small scale gardening techniques. These techniques were brought to the United States by Alan Chadwick in the 1930s. This method has since been promoted by John Jeavons, Director of Ecology Action. A garden is more than just a means of providing food, it is a model of what is possible in a community - everyone could have a garden of some kind (container, growing box, raised bed) and produce healthy, nutritious organic food, a farmers market, a place to pass on gardening experience, and a sharing of bounty, promoting a more sustainable way of living that would encourage their local economy. A simple 4' x 8' (32 square feet) raised bed garden based on the principles of bio-in-tensive planting and square foot gardening uses fewer nutrients and less water, and could keep a family, or community, supplied with an abundance of healthy, nutritious organic greens, while promoting a more sustainable way of living.

Organic gardening is designed to work with the ecological systems and minimally dis-turb the Earth's natural balance. Because of this organic farmers have been interested in reduced-tillage methods. Conventional agriculture uses mechanical tillage, which is plowing or sowing, which is harmful to the environment. The impact of tilling in organic farming is much less of an issue. Ploughing speeds up erosion because the soil remains uncovered for a long period of time and if it has a low content of organic matter the structural stability of the soil decreases. Organic farmers use techniques such as mulching, planting cover crops, and intercropping, to maintain a soil cover throughout most of the year. The use of compost, manure mulch and other organic fertilizers yields a higher organic content of soils on organic farms and helps limit soil degradation and erosion.

Other methods can also be used to supplement an existing garden. Methods such as composting, or vermicomposting. These practices are ways of recycling organic matter into some of the best organic fertilizers and soil conditioner. Vermicompost is especially easy. The byproduct is also an excellent source of nutrients for an organic garden.

Pest Control Approaches

Differing approaches to pest control are equally notable. In chemical horticulture, a specific insecticide may be applied to quickly kill off a particular insect pest. Chemical controls can dramatically reduce pest populations in the short term, yet by unavoidably killing (or starving) natural control insects and animals, cause an increase in the pest population in the long term, thereby creating an ever increasing problem. Repeated use of insecticides and herbicides also encourages rapid natural selection of resistant insects, plants and other organisms, necessitating increased use, or requiring new, more powerful controls.

In contrast, organic horticulture tends to tolerate some pest populations while taking the long view. Organic pest control requires a thorough understanding of pest life cycles and interactions, and involves the cumulative effect of many techniques, including:
• Allowing for an acceptable level of pest damage
• Encouraging predatory beneficial insects to flourish and eat pests
• Encouraging beneficial microorganisms
• Careful plant selection, choosing disease-resistant varieties
• Planting companion crops that discourage or divert pests
• Using row covers to protect crop plants during pest migration periods
• Rotating crops to different locations from year to year to interrupt pest reproduction cycles
• Using insect traps to monitor and control insect populations

Each of these techniques also provides other benefits, such as soil protection and improvement, fertilization, pollination, water conservation and season extension. These benefits are both complementary and cumulative in overall effect on site health. Organic pest control and biological pest control can be used as part of integrated pest management (IPM). However, IPM can include the use of chemical pesticides that are not part of organic or biological techniques.

Impact on the Global Food Supply

One controversy associated with organic food production is the matter of food produced per acre. Even with good organic practices, organic agriculture may be five to twenty-five percent less productive than conventional agriculture, depending on the crop.

Much of the productivity advantage of conventional agriculture is associated with the use of nitrogen fertilizer. However, the use, and especially the overuse, of nitrogen fertilizer has negative effects such as nitrogen runoff harming natural water supplies and increased global warming.

Organic methods have other advantages, such as healthier soil, that may make organic farming more resilient, and therefore more reliable in producing food, in the face of challenges such as climate change.

As well, world hunger is not primarily an issue of agricultural yields, but distribution and waste.

Kitchen Garden

The traditional kitchen garden, also known as a potager (in French, jardin potager) or in Scotland a kailyaird, is a space separate from the rest of the residential garden – the ornamental plants and lawn areas. Most vegetable gardens are still miniature versions of old family farm plots, but the kitchen garden is different not only in its history, but also its design.

The kitchen garden may serve as the central feature of an ornamental, all-season landscape, or it may be little more than a humble vegetable plot. It is a source of herbs, vegetables and fruits, but it is often also a structured garden space with a design based on repetitive geometric patterns.

The kitchen garden has year-round visual appeal and can incorporate permanent perennials or woody shrub plantings around (or among) the annuals.

Potager Garden

A potager is a French term for an ornamental vegetable or kitchen garden. The historical design precedent is from the Gardens of the French Renaissance and Baroque Garden à la française eras. Often flowers (edible and non-edible) and herbs are planted with the vegetables to enhance the garden's beauty. The goal is to make the function of providing food aesthetically joyful.

Plants are chosen as much for their functionality as for their color and form. Many are trained to grow upward. A well-designed potager can provide food as well as cut flowers and herbs for the home with very little maintenance. Potagers can disguise their function of providing for a home in a wide array of forms—from the carefree style of the cottage garden to the formality of a knot garden.

Vegetable Garden

A small vegetable garden in May outside of Austin, Texas

A vegetable garden (also known as a vegetable patch or vegetable plot) is a garden that exists to grow vegetables and other plants useful for human consumption, in contrast to a flower garden that exists for aesthetic purposes. It is a small-scale form of vegetable growing. A vegetable garden typically includes a compost heap, and several plots or divided areas of land, intended to grow one or two types of plant in each plot. Plots may also be divided into rows with an assortment of vegetables grown in the different rows. It is usually located to the rear of a property in the back garden or back yard. Many families have home kitchen and vegetable gardens that they use to produce food. In World War II, many people had a garden called a "victory garden" which provided food and thus freed resources for the war effort.

An herbal garden at Lippensgoed-Bulskampveld, Beernem, Belgium

With worsening economic conditions and increased interest in organic and sustainable living, many people are turning to vegetable gardening as a supplement to their family's diet. Food grown in the back yard consumes little if any fuel for shipping or maintenance, and the grower can be sure of what exactly was used to grow it. Organic horticulture, or organic gardening, has become increasingly popular for the modern home gardener.

Borage is commonly grown in herb gardens; its flowers can be used as a garnish

There are many types of vegetable gardens. The potager, a garden in which vegetables, herbs and flowers are grown together, has become more popular than the more traditional rows or blocks.

Cowbridge Physic Garden, Wales

Herb Garden

The herb garden is often a separate space in the garden, devoted to growing a specific group of plants known as herbs. These gardens may be informal patches of plants, or they may be carefully designed, even to the point of arranging and clipping the plants to form specific patterns, as in a knot garden.

Herb gardens may be purely functional or they may include a blend of functional and ornamental plants. The herbs are usually used to flavour food in cooking, though they may also be used in other ways, such as discouraging pests, providing pleasant scents, or serving medicinal purposes (such as a physic garden or medicinal herb garden), among others.

A kitchen garden can be created by planting different herbs in pots or containers, with the added benefit of mobility. Although not all herbs thrive in pots or containers, some herbs do better than others. Mint, a fragrant yet invasive herb, is an example of an herb that is advisable to keep in a container or it will take over the whole garden.

Some popular culinary herbs in temperate climates are to a large extent still the same as in the medieval period.

Herbs often have multiple uses. For example, mint may be used for cooking, tea, and pest control. Examples of herbs and their uses (not intended to be complete):

- Annual culinary herbs: basil, dill, summer savory

- Perennial culinary herbs: mint, rosemary, thyme, tarragon

- Herbs used for potpourri: lavender, lemon verbena

- Herbs used for tea: mint, lemon verbena, chamomile, bergamot, hibiscus

- Herbs used for other purposes: stevia for sweetening, feverfew for pest control in the garden.

Witches' Garden

A witches' garden is an herb garden specifically designed and used for the cultivation of herbs, for culinary, medicinal and/or spiritual purposes. Herbal baths, the making of incense, tied in bundles for rituals or prayers, or placed in charms are just some of the ways herbs can be used for spiritual purposes.

Herb gardens developed from the general gardens of the ancient classical world, which were used for growing vegetables, flowers, fruits and medicines. For centuries "wise women" and "healers" understood the uses of herbs for the purposes of healing and magic. During the medieval period, monks and nuns acquired this medical knowledge and grew the necessary herbs in specialized gardens.

Typical plants found within a witches' garden are the following: rosemary, sage, parsley, mint, catnip, henbane, marjoram, thyme, rue, angelica, bay, oregano, dill, aloe, arnica, chives, and basil. Basil is especially common in these gardens, not just for its culinary use, but as a strong protection herb. It is said, "Where basil grows, no evil goes!" and "Where basil is, no evil lives!" With the advance of medical and botanical sciences in Renaissance Europe, monastic herb gardens developed into botanical gardens. However, these are just examples of the common herbs found within a witch's garden. Many other plants and herbs can be grown. It is a very personal garden, and therefore is unique to the individual witch. For a "true" witch or pagan, this garden is not just used for the purposes of each of the herbs grown, but it is also a way to become in touch with mother nature and become one with the Earth.

Slow Gardening

Slow gardening is a philosophical approach to gardening which encourages participants to savor everything they do, using all the senses, through all seasons, regardless of garden type of style. Slow Gardening applies equally to people growing vegetables, herbs, flowers, and fruits, as well as those who tend to their own lawn, or have an intense garden hobby such as topiary, bonsai, or plant hybridizing. It actively promotes self-awareness, personal responsibility, and environmental stewardship.

Slow gardening, which is an attitude, not a "how-to" checklist of things to do or not do, was started by American horticulturist and garden author Felder Rushing, who was inspired by the Slow Food organization. The Slow food movement unleashed a worldwide wave of relief among people of all walks of life. Slow gardening has also been promoted by horticulturist and Mississippi radio show host, Felder Rushing.

The Slow Gardening approach can help us enjoy our gardens year in and year out while connecting us with our neighbors. It strikes a special chord among gardeners who, though perfectly normal in all respects, have struggled to find – and follow – their bliss against the lockstep pressures of "fitting in."

The basic tenets of slow gardening are rooted in the Gestalt approach. A major goal of all Slow movements is for adherents to become aware of what and how they are doing something while valuing how it affects the whole. The slow gardening concept:

- uses an experiential, hands-on approach to gardening
- takes into account the whole garden (or gardener – body, mind and spirit)
- assesses what is happening in the present (the here-and-now)
- emphasizes self-awareness
- encourages personal (garden) responsibility
- acknowledges the integrity, sensitivity, and creativity of the gardener
- recognizes that the gardener is central to the gardening process.

Companion Planting

Companion planting of carrots and onions

Companion planting in gardening and agriculture is the planting of different crops in proximity for pest control, pollination, providing habitat for beneficial creatures, maximizing use of space, and to otherwise increase crop productivity. Companion planting is a form of polyculture.

Companion planting is used by farmers and gardeners in both industrialized and developing countries for many reasons. Many of the modern principles of companion planting were present many centuries ago in cottage gardens in England and forest gardens in Asia, and thousands of years ago in Mesoamerica.

History

In China, mosquito ferns (Azolla spp.) have been used for at least a thousand years as companion plants for rice crops. They host a cyanobacterium that fixes nitrogen from the atmosphere, and they block light from plants that would compete with the rice.

Companion planting was practiced in various forms by the indigenous peoples of the Americas prior to the arrival of Europeans. These peoples domesticated squash 8,000 to 10,000 years ago, then maize, then common beans, forming the Three Sisters agricultural technique. The cornstalk served as a trellis for the beans to climb, and the beans fixed nitrogen, benefitting the maize.

Companion planting was widely promoted in the 1970s as part of the organic gardening movement. It was encouraged for pragmatic reasons, such as natural trellising, but mainly with the idea that different species of plant may thrive more when close together. It is also a technique frequently used in permaculture, together with mulching, polyculture, and changing of crops.

Examples of Companion Plants

Nasturtium (Tropaeolum majus) is a food plant of some caterpillars which feed primarily on members of the cabbage family (brassicas), and some gardeners claim that planting them around brassicas protects the food crops from damage, as eggs of the pests are preferentially laid on the nasturtium. This practice is called trap cropping (using alternative plants to attract pests away from a main crop). However, while many trap crops have successfully diverted pests off of focal crops in small scale greenhouse, garden and field experiments, only a small portion of these plants have been shown to reduce pest damage at larger commercial scales.

The smell of the foliage of marigolds is claimed to deter aphids from feeding on neighbouring crops. Marigolds with simple flowers also attract nectar-feeding adult hoverflies, the larvae of which are predators of aphids.

Various legume crops benefit from being commingled with a grassy nurse crop. For example, common vetch or hairy vetch is planted together with rye or winter wheat to make a good cover crop or green manure (or both).

The terms "undersowing" and "overseeding" both involve intercropping as a type of companion planting. "Undersowing" conveys the idea of sowing the second crop among the young plants of the first crop (or in between the rows, if rows are used). A connotation of understory growth is conveyed, albeit exaggerated (because the first crop is not yet a dense canopy). "Overseeding" conveys the idea of broadcasting the seeds of the second crop over the existing first crop. This is analogous to overseeding a lawn to improve the mix of grasses present.

Versions

There are a number of systems and ideas using companion planting.

Square foot gardening attempts to protect plants from many normal gardening problems by packing them as closely together as possible, which is facilitated by using companion plants, which can be closer together than normal.

Another system using companion planting is the forest garden, where companion plants are intermingled to create an actual ecosystem, emulating the interaction of up to seven levels of plants in a forest or woodland.

Organic gardening may make use of companion planting, since many synthetic means of fertilizing, weed reduction and pest control are forbidden.

Host-finding Disruption

Recent studies on host-plant finding have shown that flying pests are far less successful if their host-plants are surrounded by any other plant or even "decoy-plants" made of green plastic, cardboard, or any other green material.

The host-plant finding process occurs in phases:

- The first phase is stimulation by odours characteristic to the host-plant. This induces the insect to try to land on the plant it seeks. But insects avoid landing on brown (bare) soil. So if only the host-plant is present, the insects will quasi-systematically find it by simply landing on the only green thing around. This is called (from the point of view of the insect) "appropriate landing". When it does an "inappropriate landing", it flies off to any other nearby patch of green. It eventually leaves the area if there are too many 'inappropriate' landings.

- The second phase of host-plant finding is for the insect to make short flights from leaf to leaf to assess the plant's overall suitability. The number of leaf-to-leaf flights varies according to the insect species and to the host-plant stimulus received from each leaf. The insect must accumulate sufficient stimuli from the host-plant to lay eggs; so it must make a certain number of consecutive 'appropriate' landings. Hence if it makes an 'inappropriate landing', the assessment of that plant is negative, and the insect must start the process anew.

Thus it was shown that clover used as a ground cover had the same disruptive effect on eight pest species from four different insect orders. An experiment showed that 36% of cabbage root flies laid eggs beside cabbages growing in bare soil (which resulted in no crop), compared to only 7% beside cabbages growing in clover (which allowed a good crop). Simple decoys made of green cardboard also disrupted appropriate landings just as well as did the live ground cover.

Companion Plant Categories

The use of companion planting can be of benefit to the grower in a number of different ways, including:

- Hedged investment – the growing of different crops in the same space increases the odds of some yield being given, even if one crop fails.

- Increased level interaction – when crops are grown on different levels in the same space, such as providing ground cover or one crop working as a trellis for another, the overall yield of a plot may be increased.

- Protective shelter is when one type of plant may serve as a wind break or provide shade for another.

- Pest suppression – some companion plants may help prevent pest insects or pathogenic fungi from damaging the crop, through chemical means.

- Predator recruitment and positive hosting – The use of companion plants that produce copious nectar or pollen in a vegetable garden (insectary plants) may help encourage higher populations of beneficial insects that control pests, as some beneficial predatory insects only consume pests in their larval form and are nectar or pollen feeders in their adult form.

- Trap cropping – some companion plants are claimed to attract pests away from others.

- Pattern disruption – in a monoculture pests spread easily from one crop plant to the next, whereas such easy progress may be disrupted by surrounding companion plants of a different type.

List of Companion Plants

Allium

Allium is a genus of monocotyledonous flowering plants that includes the cultivated onion, garlic, scallion, shallot and leek as well as chives and hundreds of other wild species.

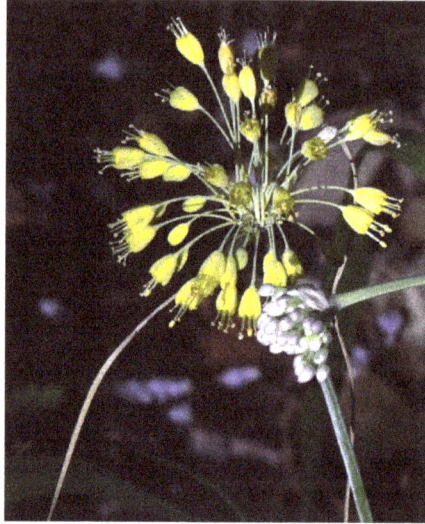

Allium flavum (yellow) and Allium carinatum (purple)

The generic name Allium is the Latin word for garlic, and Linnaeus first described the genus Allium in 1753. The cooking and consumption of parts of the plants is due to the large variety of flavours and textures of the species. Various Allium have been cultivated from the earliest times and about a dozen species are economically important as crops, or garden vegetables, and an increasing number of species are important as ornamental plants. The inclusion of a species to the genus Allium is taxonomically difficult and species boundaries are unclear. Estimates of the number of species have been as low as 260, and as high as 979, but are most likely about 800–850. The type species for the genus is Allium sativum.

Allium species occur in temperate climates of the Northern Hemisphere, except for a few species occurring in Chile (such as A. juncifolium), Brazil (A. sellovianum), and tropical Africa (A. spathaceum). They vary in height between 5 cm and 150 cm. The flowers form an umbel at the top of a leafless stalk. The bulbs vary in size between species, from small (around 2–3 mm in diameter) to rather large (8–10 cm). Some species (such as Welsh onion A. fistulosum) develop thickened leaf-bases rather than forming bulbs as such.

Plants of the Allium genus produce chemical compounds (mostly derived from cysteine sulfoxides) that give them a characteristic (alliaceous) onion or garlic taste and odor. Many are used as food plants, though not all members of the genus are equally flavorous. In most cases, both bulb and leaves are edible and the taste may be strong or weak, depending on the species and on ground sulfur (usually as sulfate) content. In the rare occurrence of sulfur-free growth conditions, all Allium species lose their usual pungency altogether.

In the APG III classification system, Allium is placed in the family Amaryllidaceae, subfamily Allioideae (formerly the family Alliaceae). In some of the older classification systems, Allium was placed in Liliaceae. Molecular phylogenetic studies have shown this circumscription of Liliaceae is not monophyletic.

Allium is one of about fifty-seven genera of flowering plants with more than 500 species. It is by far the largest genus in the Amaryllidaceae, and also in the Alliaceae in classification systems in which that family is recognized as separate.

Description

Capsule of Allium oreophilum.

Allium species are herbaceous perennials with flowers produced on scapes. They grow from solitary or clustered tunicate bulbs and many have an onion odor and taste. Plants are perennialized by bulbs that reform annually from the base of the old bulb, or are produced on the ends of rhizomes or, in a few species, at the ends of stolons. A small number of species have tuberous roots. The bulbs' outer coats are commonly brown or grey, with a smooth texture, and are fibrous, or with cellular reticulation. The inner coats of the bulbs are membranous.

Many alliums have basal leaves that commonly wither away from the tips downward before or while the plants flower, but some species have persistent foliage. Plants produce from one to 12 leaves, most species having linear, channeled or flat leaf blades. The leaf blades are straight or variously coiled, but some species have broad leaves, including A. victorialis and A. tricoccum. The leaves are sessile, and very rarely narrowed into a petiole.

The flowers are erect or in some species pendent, having six petal-like tepals produced in two whorls. The flowers have one style and six epipetalous stamens; the anthers and pollen can vary in color depending on the species. The ovaries are superior, and three-lobed with three locules.

The fruits are capsules that open longitudinally along the capsule wall between the partitions of the locule. The seeds are black, and have a rounded shape.

The terete or flattened flowering scapes are normally persistent. The inflorescences are umbels, in which the outside flowers bloom first and flowering progresses to the inside. Some species produce bulbils within the umbels, and in some species, such as Allium paradoxum, the bulbils replace some or all the flowers. The umbels are subtended by noticeable spathe bracts, which are commonly fused and normally have around three veins.

Some bulbous alliums increase by forming little bulbs or "offsets" around the old one, as well as by seed. Several species can form many bulbils in the flowerhead; in the so-called "tree onion" or Egyptian onion (A. × proliferum) the bulbils are few, but large enough to be pickled.

Many of the species of Allium have been used as food items throughout their ranges. There are several poisonous species that are somewhat similar in appearance (e.g. in North America, death camas, Toxicoscordion venenosum), but none of these has the distinctive scent of onions or garlic.

Taxonomy

With over 850 species Allium is the sole genus in the Allieae, one of four tribes of subfamily Allioideae (Amaryllidaceae). New species continue to be described and Allium is one of the largest monocotyledonous genera, but the precise taxonomy of Allium is poorly understood, with incorrect descriptions being widespread. The difficulties arise from the fact that the genus displays considerable polymorphism and has adapted to a wide variety of habitats. Furthermore, traditional classications had been based on homoplasious characteristics (the independent evolution of similar features in species of different lineages). However, the genus has been shown to be monophyletic, containing three major clades, although some proposed subgenera are not. Some progress is being made using molecular phylogenetic methods, and the internal transcribed spacer (ITS) region, including the 5.8S rDNA and the two spacers ITS1 and ITS2, is one of the more commonly used markers in the study of the differentiation of the Allium species.

Allium includes a number of taxonomic groupings previously considered separate genera (Caloscordum Herb., Milula Prain and Nectaroscordum Lindl.) Allium spicatum had been treated by many authors as Milula spicata, the only species in the monospecific genus Milula. In 2000, it was shown to be embedded in Allium.

Phylogeny

History

When Linnaeus formerly described the genus Allium in his Species Plantarum (1753), there were thirty species with this name. He placed Allium in a grouping he referred to as Hexandria monogynia (i.e. six stamens and one pistil) containing 51 genera in all.

Subdivision

Linnaeus originally grouped his 30 species into three alliances, e.g. Foliis caulinis planis. Since then, many attempts have been made to divide the growing number of recognised species into infrageneric subgroupings, initially as sections, and then as subgenera further divided into sections. The modern era of phylogenetic analysis dates to 1996. In 2006 Friesen, Fritsch, and Blattner described a new classification with 15 subgenera, 56 sections, and about 780 species based on the nuclear ribosomal gene internal transcribed spacers. Some of the subgenera correspond to the once separate genera (Caloscordum, Milula, Nectaroscordum) included in the Gilliesieae. The terminology has varied with some authors subdividing subgenera into Sections and others Alliances. The term Alliance has also been used for subgroupings within species, e.g. Allium nigrum, and for subsections.

Subsequent molecular phylogenetic studies have shown the 2006 classification is a considerable improvement over previous classifications, but some of its subgenera and sections are probably not monophyletic. Meanwhile the number of new species continued to increase, reaching 800 by 2009, and the pace of discovery has not decreased. Detailed studies have focused on a number of subgenera, including Amerallium. Amerallium is strongly supported as monophyletic. Subgenus Melanocrommyum has also been the subject of considerable study, while work on subgenus Allium has focussed on section Allium, including Allium ampeloprasum, although sampling was not sufficient to test the monophyly of the section.

The major evolutionary lineages or lines correspond to the three major clades. Line one (the oldest) with three subgenera is predominantly bulbous, the second, with five subgenera and the third with seven subgenera contain both bulbous and rhizomatous taxa.

Evolutionary Lines and Subgenera

The three evolutionary lineages and 15 subgenera represent the classification scheme of Friesen et al. (2006) and Li (2010). (number of sections/number of species)

- First evolutionary line

    1. Nectaroscordum (Lindl.) Asch. et Graebn Type: Allium siculum (1/3) Mediterranean bells, Sicilian honey garlic

    2. Microscordum (Maxim.) N. Friesen Type: Allium monanthum (1/1)

    3. Amerallium Traub Type: Allium canadense (12/135)

- Second evolutionary line

1. Caloscordum (Herb.) R. M. Fritsch Type: Allium neriniflorum (1/3)

2. Anguinum (G. Don ex Koch) N. Friesen Type: Allium victorialis (1/12)

3. Porphyroprason (Ekberg) R. M. Fritsch Type: Allium oreophilum (1/1)

4. Vvedenskya (Kamelin) R. M. Fritsch Type: Allium kujukense (1/1)

5. Melanocrommyum (Webb et Berthel.) Rouy Type: Allium nigrum (20/160)

- Third evolutionary line

  1. Butomissa (Salisb.) N. Friesen Type: Allium ramosum (2/4) fragrant garlic

  2. Cyathophora R. M. Fritsch Type: Allium cyathophorum (3/5)

  3. Rhizirideum (G. Don ex Koch) Wendelbo s.s Type: Allium senescens (5/37)

  4. Allium L. Type: Allium sativum (15/300)

  5. Reticulatobulbosa (Kamelin) N. Friesen Type: Allium lineare (5/80)

  6. Polyprason Radic Type: Allium moschatum (4/50)

  7. Cepa (Mill.) Radic ´ Type: Allium cepa (5/30) onion, garden onion, bulb onion, common onion

First Evolutionary Line

Although this lineage consists of three subgenera, nearly all the species are attributed to subgenus Amerallium, the third largest subgenus of Allium. The lineage is considered to represent the most ancient line within Allium, and to be the only lineage that is purely bulbous, the other two having both bulbous and rhizomatous taxa. Within the lineage Amerallium is a sister group to the other two subgenera (Microscordum+Nectaroscordum).

Second Evolutionary Line

Nearly all the species in this lineage of five subgenera are accounted for by subgenus Melanocrommyum, which is most closely associated with subgenera Vvedenskya and Porphyroprason, phylogenetically. These three genera are late-branching whereas the remaining two subgenera, Caloscordum and Anguinum are early branching.

Third Evolutionary Line

The third evolutionary line contains the most number of sections (seven) and also the largest subgenus of the Allium genus, subgenus Allium which includes the type species of the genus, Allium sativum. This subgenus also contains the majority of the species in the line. Within the lineage the phylogeny is complex. Two small subgenera Butomissa and Cyathophora form a sister clade to the remaining five subgenera, with Butomissa as the first branching group. Amongst the remaining five subgenera, Rhizirideum forms a medium-sized subgenus that is the sister to the other four larger subgenera. However, they may not be monophyletic.

Distribution and Habitat

The majority of Allium species are native to the Northern Hemisphere, being spread throughout the holarctic region, from dry subtropics to the boreal zone, but predominantly in Asia. Of the latter 138 species occur in China, about a sixth of all species, representing five subgenera. A few species are native to Africa and Central and South America. A single known exception, Allium dregeanum occurs in the Southern Hemisphere (South Africa). There are two centres of diversity, a major one from the Mediterranean Basin to Central Asia and Pakistan, while a minor one is found in western North America.

Ecology

Species grow in various conditions from dry, well-drained mineral-based soils to moist, organic soils; most grow in sunny locations, but a number also grow in forests (e.g., A. ursinum), or even in swamps or water.

Various Allium species are used as food plants by the larvae of the leek moth and onion fly as well as some Lepidoptera including cabbage moth, common swift moth (recorded on garlic), garden dart moth, large yellow underwing moth, nutmeg moth, setaceous Hebrew character moth, turnip moth and Schinia rosea, a moth that feeds exclusively on Allium species.

Cultivation

Selection of cultivated alliums displayed at the BBC Gardeners' World show

Many Allium species have been harvested through human history, but only about a dozen are still economically important today as crops or garden vegetables.

Ornamental

Many Allium species and hybrids are cultivated as ornamentals. These include A. cristophii and A. giganteum, which are used as border plants for their ornamental flowers, and their "architectural" qualities. Several hybrids have been bred, or selected, with rich purple flowers. A. hollandicum 'Purple Sensation' is one of the most popular and has been given an Award of Garden Merit (H4). These ornamental onions produce spherical umbels on single stalks in spring and summer, in a wide variety of sizes and colours, ranging from white (Allium 'Mont Blanc'), blue (A. caeruleum), to yellow (A. flavum) and purple (A. giganteum). By contrast, other species (such as invasive A. triquetrum and A. ursinum) can become troublesome garden weeds. The hybrid cultivars 'Beau Regard', 'Gladiator', and 'Globemaster' have gained the Royal Horticultural Society's Award of Garden Merit.

Toxicity

Dogs and cats are very susceptible to poisoning after the consumption of certain species.

Uses

The genus includes many economically important species. These include onions (A. cepa), French shallots (A. oschaninii), leeks (A. ampeloprasum), scallions (various Allium species), and herbs such as garlic (A. sativum) and chives (A. schoenoprasum). Some have been used as traditional medicines.

Legume

A selection of various legumes

Legumes are grown agriculturally, primarily for their grain seed called pulse, for livestock forage and silage, and as soil-enhancing green manure. Well-known legumes include alfalfa, clover, peas, beans, lentils, lupin bean, mesquite, carob, soybeans, peanuts, and tamarind.

A legume fruit is a simple dry fruit that develops from a simple carpel and usually dehisces (opens along a seam) on two sides. A common name for this type of fruit is a pod, although the term "pod" is also applied to a few other fruit types, such as that of vanilla (a capsule) and of radish (a silique).

Legumes are notable in that most of them have symbiotic nitrogen-fixing bacteria in structures called root nodules. For that reason, they play a key role in crop rotation.

Terminology

The term "pulse", as used by the United Nations' Food and Agriculture Organization (FAO), is reserved for crops harvested solely for the dry seed. This excludes green beans and green peas, which are considered vegetable crops. Also excluded are seeds that are mainly grown for oil extraction (oilseeds like soybeans and peanuts), and seeds which are used exclusively for sowing forage (clovers, alfalfa). However, in common usage, these distinctions are not always clearly made, and many of the varieties used for dried pulses are also used for green vegetables, with their beans in pods while young.

Some Fabaceae, such as Scotch broom and other Genisteae, are leguminous but are usually not called legumes by farmers, who tend to restrict that term to food crops.

Uses

Farmed legumes can belong to many agricultural classes, including forage, grain, blooms, pharmaceutical/industrial, fallow/green manure, and timber species. Most commercially farmed species fill two or more roles simultaneously, depending upon their degree of maturity when harvested.

Human Consumption

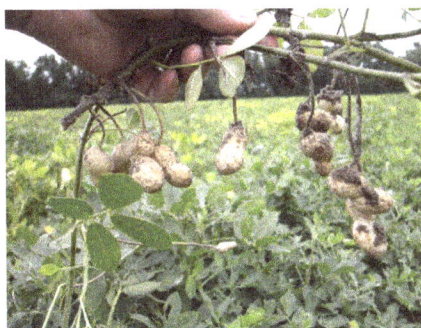

Freshly dug peanuts (Arachis hypogaea), indehiscent legume fruits

Grain legumes are cultivated for their seeds. The seeds are used for human and animal consumption or for the production of oils for industrial uses. Grain legumes include beans, lentils, lupins, peas, and peanuts.

Nutritional Value

Legumes are a significant source of protein, dietary fiber, carbohydrates and dietary minerals; for example, a 100 gram serving of cooked chickpeas contains 18% of the Daily Value (DV) for protein, 30% DV for dietary fiber, 43% DV for folate and 52% DV for manganese. Like other plant-based foods, pulses contain no cholesterol and little fat or sodium.

Legumes are also an excellent source of resistant starch which is broken down by bacteria in the large intestine to produce short-chain fatty acids used by intestinal cells for food energy.

Preliminary studies in humans include the potential for regular consumption of legumes in a vegetarian diet to affect metabolic syndrome. There is evidence that a portion of pulses (roughly one cup daily) in a diet may help lower blood pressure and reduce LDL cholesterol levels, though there is a concern about the quality of the supporting data.

Classification

Depending on the variety, Phaseolus vulgaris (a pulse) may be called "common bean", "kidney bean", "haricot bean", "pinto bean", "navy bean", among other names.

FAO recognizes 11 primary pulses.

- Dry beans (Phaseolus spp. including several species now in Vigna)
    - Kidney bean, navy bean, pinto bean, haricot bean (Phaseolus vulgaris)
    - Lima bean, butter bean (Phaseolus lunatus)
    - Adzuki bean, azuki bean (Vigna angularis)
    - Mung bean, golden gram, green gram (Vigna radiata)
    - Black gram, urad (Vigna mungo)

- Scarlet runner bean (Phaseolus coccineus)

- Ricebean (Vigna umbellata)

- Moth bean (Vigna aconitifolia)

- Tepary bean (Phaseolus acutifolius)

- Dry broad beans (Vicia faba)

  - Horse bean (Vicia faba equina)

  - Broad bean (Vicia faba)

  - Field bean (Vicia faba)

- Dry peas (Pisum spp.)

  - Garden pea (Pisum sativum var. sativum)

  - Protein pea (Pisum sativum var. arvense)

- Chickpea, garbanzo, Bengal gram (Cicer arietinum)

- Dry cowpea, black-eyed pea, blackeye bean (Vigna unguiculata)

- Pigeon pea, Arhar/Toor, cajan pea, Congo bean, gandules (Cajanus cajan)

- Lentil (Lens culinaris)

- Bambara groundnut, earth pea (Vigna subterranea)

- Vetch, common vetch (Vicia sativa)

- Lupins (Lupinus spp.)

- Minor pulses, including:

  - Lablab, hyacinth bean (Lablab purpureus)

  - Jack bean (Canavalia ensiformis), sword bean (Canavalia gladiata)

  - Winged bean (Psophocarpus tetragonolobus)

  - Velvet bean, cowitch (Mucuna pruriens var. utilis)

  - Yam bean (Pachyrhizus erosus)

Forage

Forage legumes are of two broad types. Some, like alfalfa, clover, vetch (Vicia), stylo (Stylosanthes), or Arachis, are sown in pasture and grazed by livestock. Other forage legumes such as Leucaena or Albizia are woody shrub or tree species that are either broken down by livestock or regularly cut by humans to provide livestock feed.

White clover, a forage crop

Other uses

Lupin flower garden

Legume species grown for their flowers include lupins, which are farmed commercially for their blooms as well as being popular in gardens worldwide. Industrially farmed legumes include Indigofera and Acacia species, which are cultivated for dye and natural gum production, respectively. Fallow/green manure legume species are cultivated to be tilled back into the soil in order to exploit the high levels of captured atmospheric nitrogen found in the roots of most legumes. Numerous legumes farmed for this purpose include Leucaena, Cyamopsis, and Sesbania species. Various legume species are farmed for timber production worldwide, including numerous Acacia species and Castanospermum australe.

Legume trees like the locust trees (Gleditsia, Robinia) or the Kentucky coffeetree (Gymnocladus dioicus) can be used in permaculture food forests. Other legume trees like laburnum and the woody climbing vine wisteria are poisonous.

Nitrogen Fixation

Root nodules on a Wisteria plant (a hazelnut pictured for comparison)

Many legumes contain symbiotic bacteria called Rhizobia within root nodules of their root systems. (Plants belonging to the genus Styphnolobium are one exception to this rule.) These bacteria have the special ability of fixing nitrogen from atmospheric, molecular nitrogen (N_2) into ammonia (NH_3). The chemical reaction is:

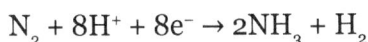

$$N_2 + 8H^+ + 8e^- \rightarrow 2NH_3 + H_2$$

Ammonia is then converted to another form, ammonium (NH_4^+), sable by (some) plants by the following reaction:

$$NH_3 + H^+ \rightarrow NH_4^+$$

This arrangement means that the root nodules are sources of nitrogen for legumes, making them relatively rich in plant proteins. All proteins contain nitrogenous amino acids. Nitrogen is therefore a necessary ingredient in the production of proteins. Hence, legumes are among the best sources of plant protein.

When a legume plant dies in the field, for example following the harvest, all of its remaining nitrogen, incorporated into amino acids inside the remaining plant parts, is released back into the soil. In the soil, the amino acids are converted to nitrate ($NO-3$), making the nitrogen available to other plants, thereby serving as fertilizer for future crops.

In many traditional and organic farming practices, crop rotation involving legumes is common. By alternating between legumes and nonlegumes, sometimes planting nonlegumes two times in a row and then a legume, the field usually receives a sufficient amount of nitrogenous compounds to produce a good result, even when the crop is nonleguminous. Legumes are sometimes referred to as "green manure".

History

Archaeologists have discovered traces of pulse production around Ravi River (Punjab), the seat of the Indus Valley Civilisation, dating circa 3300 BC. Meanwhile, evidence of lentil cultivation has also been found in Egyptian pyramids and dry pea seeds have been discovered in a Swiss village that are believed to date back to the Stone Age. Archaeological evidence suggests that these peas must have been grown in the eastern Mediterranean and Mesopotamia regions at least 5,000 years ago and in Britain as early as the 11th century.

World Economy

India is the world's largest producer and the largest consumer of pulses. Pakistan, Canada, Myanmar, Australia and the United States, in that order, are significant exporters and are India's most significant suppliers. The global pulse market is estimated at 60 million tonnes.

International Year of Pulses

The International Year of Pulses 2016 (IYP 2016) was declared by the Sixty-eighth session of the United Nations General Assembly. The Food and Agriculture Organization of the United Nations has been nominated to facilitate the implementation of IYP 2016 in collaboration with governments, relevant organizations, non-governmental organizations and other relevant stakeholders. Its aim is to heighten public awareness of the nutritional benefits of pulses as part of sustainable food production aimed towards food security and nutrition. IYP 2016 will create an opportunity to encourage connections throughout the food chain that would better utilize pulse-based proteins, further global production of pulses, better utilize crop rotations and address challenges in the global trade of pulses.

Logo of International Year of Pulses 2016

Beetroot

The beetroot is the taproot portion of the beet plant, usually known in North America as the beet, also table beet, garden beet, red beet, or golden beet. It is one of several of the cultivated varieties of Beta vulgaris grown for their edible taproots and their leaves (called beet greens). These varieties have been classified as B. vulgaris subsp. vulgaris Conditiva Group.

Other than as a food, beets have use as a food colouring and as a medicinal plant. Many beet products are made from other Beta vulgaris varieties, particularly sugar beet.

Uses

Food

A bundle of beetroot

Usually the deep purple roots of beetroot are eaten boiled, roasted or raw, and either alone or combined with any salad vegetable. A large proportion of the commercial production is processed into boiled and sterilized beets or into pickles. In Eastern Europe, beet soup, such as borscht, is a popular dish. In Indian cuisine, chopped, cooked, spiced beet is a common side dish. Yellow-coloured beetroots are grown on a very small scale for home consumption.

Section through taproot

The green, leafy portion of the beet is also edible. The young leaves can be added raw to salads, whilst the adult leaves are most commonly served boiled or steamed, in which case they have a taste and texture similar to spinach. Those greens selected should be from bulbs that are unmarked, instead of those with overly limp leaves or wrinkled skins, both of which are signs of dehydration. The domestication of beets can be traced to the emergence of an allele which enables biennial harvesting of leaves and taproot.

Yellow beetroot

Beetroot can be: boiled or steamed, peeled and then eaten warm with or without butter as a delicacy; cooked, pickled, and then eaten cold as a condiment; or peeled, shredded raw, and then eaten as a salad. Pickled beets are a traditional food in many countries.

A traditional Pennsylvania Dutch dish is pickled beet egg. Hard-boiled eggs are refrigerated in the liquid left over from pickling beets and allowed to marinate until the eggs turn a deep pink-red colour.

In Poland and Ukraine, beet is combined with horseradish to form popular ćwikła, which is traditionally used with cold cuts and sandwiches, but often also added to a meal consisting of meat and potatoes. The same in Serbia where the popular cvekla is used as winter salad, seasoned with salt and vinegar, with meat dishes. As an addition to horseradish it is also used to produce the "red" variety of chrain, a popular condiment in Ashkenazi Jewish, Hungarian, Polish, Russian and Ukrainian cuisine.

Popular in Australian hamburgers, a slice of pickled beetroot is combined with fried egg and sometimes pineapple (as well as the usual beef patty, barbecue/tomato sauce and salad) to make an Aussie burger.

When beet juice is used, it is most stable in foods with a low water content, such as frozen novelties and fruit fillings. Betanins, obtained from the roots, are used industrially as red food colourants, e.g. to intensify the colour of tomato paste, sauces, desserts, jams and jellies, ice cream, sweets, and breakfast cereals.

Beetroot can also be used to make wine.

Food shortages in Europe following World War I caused great hardships, including cases of mangelwurzel disease, as relief workers called it. It was symptomatic of eating only beets.

- Beetroot as food

Borscht

Salad of grated beet and apple

Finnish rosolli

Red, white and golden beetroot

Other Uses

Betanin, obtained from the roots, is used industrially as red food colorant, to improve the color and flavor of tomato paste, sauces, desserts, jams and jellies, ice cream, sweets, breakfast cereals, etc.

Historical Uses

From the Middle Ages, beetroot was used as a treatment for a variety of conditions, especially illnesses relating to digestion and the blood. Bartolomeo Platina recommended taking beetroot with garlic to nullify the effects of "garlic-breath".

During the middle of the 19th century wine often was coloured with beetroot juice.

Preliminary Research

In preliminary research, beetroot juice reduced blood pressure in hypertensive individuals and so may have an effect on mechanisms of cardiovascular disease. Tentative evidence has found that dietary nitrate supplementation such as from beets and other vegetables results in a small to moderate improvement in endurance exercise performance.

Beets contain betaines which may function to reduce the concentration of homocysteine, a homolog of the naturally occurring amino acid cysteine. High circulating levels of homocysteine may be harmful to blood vessels and thus contribute to the development of cardiovascular disease. This hypothesis is controversial as it has not yet been established whether homocysteine itself is harmful or is just an indicator of increased risk for cardiovascular disease.

Nutrition

Beetroots, cooked	
Nutritional value per 100 g (3.5 oz)	
Energy	180 kJ (43 kcal)
Carbohydrates	9.96 g
Sugars	7.96 g
Dietary fiber	2.0 g
Fat	0.18 g
Protein	1.68 g
Vitamins	
Vitamin A equiv.	(0%) 2 µg
Thiamine (B1)	(3%) 0.031 mg
Riboflavin (B2)	(2%) 0.027 mg
Niacin (B3)	(2%) 0.331 mg
Pantothenic acid (B5)	(3%) 0.145 mg
Vitamin B6	(5%) 0.067 mg
Folate (B9)	(20%) 80 µg
Vitamin C	(4%) 3.6 mg
Minerals	
Calcium	(2%) 16 mg

Iron	(6%) 0.79 mg
Magnesium	(6%) 23 mg
Manganese	(14%) 0.3 mg
Phosphorus	(5%) 38 mg
Potassium	(6%) 305 mg
Sodium	(5%) 77 mg
Zinc	(4%) 0.35 mg
• Units	
• μg = micrograms • mg = milligrams	
• IU = International units	
Percentages are roughly approximated using US recommendations for adults. Source: USDA Nutrient Database	

Per 100 gram serving providing 43 calories, beetroot is an excellent source (20% of the Daily Value, DV) of folate and a good source (14% DV) of manganese, with other nutrients in low amounts.

Safety

The red colour compound betanin is not broken down in the body, and in higher concentrations may temporarily cause urine and stool to assume a reddish colour; in the case of urine this is called beeturia. This effect may cause distress and concern due to the visual similarity to what appears to be blood in the stool, hematuria (blood in the urine), or hematochezia (blood passing through the anus, usually in or with stool). These deceptive appearances are completely harmless and subside once the betanin is out of the system. In the cases of reddish feces, the bright redness from betanin is in contrast to what occurs with melena (very dark or blackish feces) which often indicates that bleeding is occurring further up the digestive system, and is more likely to be a serious problem.

Nitrosamine formation in beet juice can reliably be prevented by adding ascorbic acid.

Cultivars

Below is a list of several commonly available cultivars of beets. Generally, 55 to 65 days are needed from germination to harvest of the root. All cultivars can be harvested earlier for use as greens. Unless otherwise noted, the root colours are shades of red and dark red with different degrees of zoning noticeable in slices.

- 'Albino', heirloom (white root)

- 'Bull's Blood', heirloom

- 'Chioggia', heirloom (distinct red and white zoned root)

- 'Crosby's Egyptian', heirloom

- 'Cylindra' / 'Formanova', heirloom (elongated root)

- 'Detroit Dark Red Medium Top', heirloom

- 'Early Wonder', heirloom

- 'Golden Beet' / 'Burpee's Golden', heirloom (yellow root)

- 'Perfected Detroit', 1934 AAS winner

- 'Red Ace' Hybrid

- 'Ruby Queen', 1957 AAS winner

- 'Touchstone Gold' (yellow root)

Bean

"Painted Pony" dry bean (Phaseolus vulgaris)

Bean is a common name for large seeds of several genera of the flowering plant family Fabaceae (also known as Leguminosae) which are used for human or animal food.

Bean plant

Terminology

The word "bean" and its Germanic cognates (e.g., Bohne) have existed in common use in West Germanic languages since before the 12th century, referring to broad beans and other pod-borne seeds. This was long before the New World genus Phaseolus was known in Europe. After Columbian-era contact between Europe and the Americas, use of the word was extended to pod-borne seeds of Phaseolus, such as the common bean and the runner bean, and the related genus Vigna. The term has long been applied generally to many other seeds of similar form, such as Old World soybeans, peas, chickpeas (garbanzo beans), other vetches, and lupins, and even to those with slighter resemblances, such as coffee beans, vanilla beans, castor beans, and cocoa beans. Thus the term "bean" in general usage can mean a host of different species.

Seeds called "beans" are often included among the crops called "pulses" (legumes), although a narrower prescribed sense of "pulses" reserves the word for leguminous crops harvested for their dry grain. The term bean usually excludes legumes with tiny seeds and which are used exclusively for forage, hay, and silage purposes (such as clover and alfalfa). According to the United Nations Food and Agriculture Organization the term bean should include only species of Phaseolus; however, enforcing that prescription has proven difficult for several reasons. One is that in the past, several species, including Vigna angularis (azuki bean), mungo (black gram), radiata (green gram), and aconitifolia (moth bean), were classified as Phaseolus and later reclassified. Another is that it is not surprising that the prescription on limiting the use of the word, because it tries to replace the word's older senses with a newer one, has never been consistently followed in general usage.

Cultivation

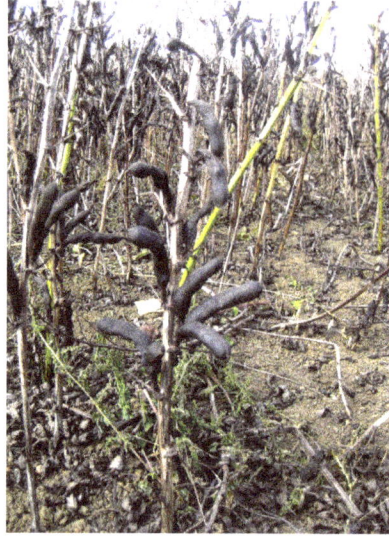

Field beans (broad beans, Vicia faba), ready for harvest

Unlike the closely related pea, beans are a summer crop that need warm temperatures to grow. Maturity is typically 55–60 days from planting to harvest. As the bean pods mature, they turn yellow and dry up, and the beans inside change from green to their mature colour. As a vine, bean plants need external support, which may be provided in the form of special "bean cages" or poles. Native Americans customarily grew them along with corn and squash (the so-called Three Sisters), with the tall cornstalks acting as support for the beans.

In more recent times, the so-called "bush bean" has been developed which does not require support and has all its pods develop simultaneously (as opposed to pole beans which develop gradually). This makes the bush bean more practical for commercial production.

History

The Beaneater (1580-90) by Annibale Carracci

Cooked beans on toast

Beans are one of the longest-cultivated plants. Broad beans, also called fava beans, in their wild state the size of a small fingernail, were gathered in Afghanistan and the Himalayan foothills. In a form improved from naturally occurring types, they were grown in Thailand since the early seventh millennium BCE, predating ceramics. They were deposited with the dead in ancient Egypt. Not until the second millennium BCE did cultivated, large-seeded broad beans appear in the Aegean, Iberia and transalpine Europe. In the Iliad (8th century BCE) is a passing mention of beans and chickpeas cast on the threshing floor.

Beans were an important source of protein throughout Old and New World history, and still are today.

The oldest-known domesticated beans in the Americas were found in Guitarrero Cave, an archaeological site in Peru, and dated to around the second millennium BCE.

Most of the kinds commonly eaten fresh or dried, those of the genus Phaseolus, come originally from the Americas, being first seen by a European when Christopher Columbus, during his exploration of what may have been the Bahamas, found them growing in fields. Five kinds of Phaseolus beans were domesticated by pre-Columbian peoples: common beans (Phaseolus vulgaris) grown from Chile to the northern part of what is now the United States, and lima and sieva beans (Phaseolus lunatus), as well as the less widely distributed teparies (Phaseolus acutifolius), scarlet runner beans (Phaseolus coccineus) and polyanthus beans (Phaseolus polyanthus) One especially famous use of beans by pre-Columbian people as far north as the Atlantic seaboard is the "Three Sisters" method of companion plant cultivation:

> In the New World, many tribes would grow beans together with maize (corn), and squash. The corn would not be planted in rows as is done by European agriculture, but in a checkerboard/hex fashion across a field, in separate patches of one to six stalks each.

> Beans would be planted around the base of the developing stalks, and would vine their way up as the stalks grew. All American beans at that time were vine plants, "bush beans" having been bred only more recently. The cornstalks would

work as a trellis for the beans, and the beans would provide much-needed nitrogen for the corn.

Squash would be planted in the spaces between the patches of corn in the field. They would be provided slight shelter from the sun by the corn, would shade the soil and reduce evaporation, and would deter many animals from attacking the corn and beans because their coarse, hairy vines and broad, stiff leaves are difficult or uncomfortable for animals such as deer and raccoons to walk through, crows to land on, etc.

Dry beans come from both Old World varieties of broad beans (fava beans) and New World varieties (kidney, black, cranberry, pinto, navy/haricot).

Beans are a heliotropic plant, meaning that the leaves tilt throughout the day to face the sun. At night, they go into a folded "sleep" position.

Types

Currently, the world genebanks hold about 40,000 bean varieties, although only a fraction are mass-produced for regular consumption.

Beans, average, canned, sugarfree	
Nutritional value per 100 g (3.5 oz)	
Energy	334 kJ (80 kcal)
Carbohydrates	10.5 g
Fat	0.5 g
Protein	9.6 g
• Units	
• μg = micrograms • mg = milligrams	
• IU = International units	
Percentages are roughly approximated using US recommendations for adults.	

Some bean types include:

- Vicia

  - Vicia faba (broad bean or fava bean)

Vicia faba or broad beans, known in the US as fava beans

- Phaseolus

    - Phaseolus acutifolius (tepary bean)

    - Phaseolus coccineus (runner bean)

    - Phaseolus lunatus (lima bean)

    - Phaseolus vulgaris (common bean; includes the pinto bean, kidney bean, black bean, Appaloosa bean as well as green beans, and many others)

    - Phaseolus polyanthus (a.k.a. P. dumosus, recognized as a separate species in 1995)

- Vigna

    - Vigna aconitifolia (moth bean)

    - Vigna angularis (adzuki bean)

    - Vigna mungo (urad bean)

    - Vigna radiata (mung bean)

    - Vigna subterranea (Bambara bean or ground-bean)

    - Vigna umbellata (ricebean)

    - Vigna unguiculata (cowpea; also includes the black-eyed pea, yardlong bean and others)

- Cicer

    - Cicer arietinum (chickpea or garbanzo bean)

- Pisum

    - Pisum sativum (pea)

- Lathyrus

- Lathyrus sativus (Indian pea)
- Lathyrus tuberosus (tuberous pea)
- Lens

Lentils

- Lens culinaris (lentil)
- Lablab
  - Lablab purpureus (hyacinth bean)

Hyacinth beans

- Glycine
  - Glycine max (soybean)
- Psophocarpus
  - Psophocarpus tetragonolobus (winged bean)

Psophocarpus tetragonolobus (winged bean)

- Cajanus
    - Cajanus cajan (pigeon pea)
- Mucuna
    - Mucuna pruriens (velvet bean)
- Cyamopsis
    - Cyamopsis tetragonoloba or (guar)
- Canavalia
    - Canavalia ensiformis (jack bean)
    - Canavalia gladiata (sword bean)
- Macrotyloma
    - Macrotyloma uniflorum (horse gram)
- Lupinus (lupin)
    - Lupinus mutabilis (tarwi)
    - Lupinus albus (lupini bean)
- Arachis
    - Arachis hypogaea (peanut)

Health Concerns

Toxins

Some kinds of raw beans contain a harmful tasteless toxin, lectin phytohaemagglutinin, that must be removed by cooking. Red and kidney beans are particularly toxic, but other types also pose risks of food poisoning. A recommended method is to boil the beans for at least ten minutes; undercooked beans may be more toxic than raw beans.

Cooking beans, without bringing them to the boil, in a slow cooker at a temperature well below boiling may not destroy toxins. A case of poisoning by butter beans used to make falafel was reported; the beans were used instead of traditional broad beans or chickpeas, soaked and ground without boiling, made into patties, and shallow fried.

Bean poisoning is not well known in the medical community, and many cases may be misdiagnosed or never reported; figures appear not to be available. In the case of the UK National Poisons Information Service, available only to health professionals, the dangers of beans other than red beans were not flagged as of 2008.

Fermentation is used in some parts of Africa to improve the nutritional value of beans by removing toxins. Inexpensive fermentation improves the nutritional impact of flour from dry beans and improves digestibility, according to research co-authored by Emire Shimelis, from the Food Engineering Program at Addis Ababa University. Beans are a major source of dietary protein in Kenya, Malawi, Tanzania, Uganda and Zambia.

Bacterial Infection from Bean Sprouts

It is common to make beansprouts by letting some types of bean, often mung beans, germinate in moist and warm conditions; beansprouts may be used as ingredients in cooked dishes, or eaten raw or lightly cooked. There have been many outbreaks of disease from bacterial contamination, often by salmonella, listeria, and Escherichia coli, of beansprouts not thoroughly cooked, some causing significant mortality.

Antinutrients

Many types of bean[specify] contain significant amounts of antinutrients that inhibit some enzyme processes in the body. Phytic acid and phytates, present in grains, nuts, seeds and beans, interfere with bone growth and interrupt vitamin D metabolism. Pioneering work on the effect of phytic acid was done by Edward Mellanby from 1939.

Nutrition

Beans have significant amounts of fiber and soluble fiber, with one cup of cooked beans providing between nine and 13 grams of fiber. Soluble fiber can help lower blood cholesterol. Beans are also high in protein, complex carbohydrates, folate, and iron.

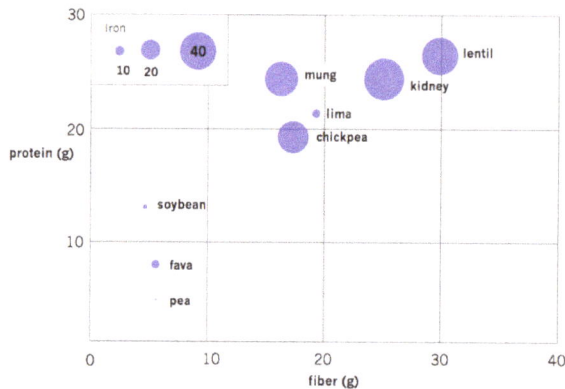

This figure shows the grams of fiber and protein per 100 gram serving of each legume. The size of the circle is proportional to its iron content. From this view, lentil and kidney beans are by far the healthiest while soybeans and peas have the least nutrients per serving.

Flatulence

Many edible beans, including broad beans and soybeans, contain oligosaccharides (particularly raffinose and stachyose), a type of sugar molecule also found in cabbage. An anti-oligosaccharide enzyme is necessary to properly digest these sugar molecules. As a normal human digestive tract does not contain any anti-oligosaccharide enzymes, consumed oligosaccharides are typically digested by bacteria in the large intestine. This digestion process produces flatulence-causing gases as a byproduct. Since sugar dissolves in water, another method of reducing flatulence associated with eating beans is to drain the water in which the beans have been cooked.

Some species of mold produce alpha-galactosidase, an anti-oligosaccharide enzyme, which humans can take to facilitate digestion of oligosaccharides in the small intestine. This enzyme, currently sold in the United States under the brand-names Beano and Gas-X Prevention, can be added to food or consumed separately. In many cuisines beans are cooked along with natural carminatives such as anise seeds, coriander seeds and cumin.

One effective strategy is to soak beans in alkaline (baking soda) water overnight before rinsing thoroughly. Sometimes vinegar is added, but only after the beans are cooked as vinegar interferes with the beans' softening.

Fermented beans will usually not produce most of the intestinal problems that unfermented beans will, since yeast can consume the offending sugars.

Production

Lablab bean and bean flower cultivated in West Bengal, India

The world leader in production of dry beans is Burma, followed by India and Brazil. In Africa, the most important producer is Tanzania.

Top ten dry bean producers—2013		
Country	Production (tonnes)	Footnote
Myanmar	3,800,000	F
India	3,630,000	
Brazil	2,936,444	A

People's Republic of China		1,400,000	*
Mexico		1,294,634	
Tanzania		1,150,000	F
United States		1,110,668	
Kenya		529,265	F
Uganda		461,000	*
Rwanda		438,236	
World		23,139,004	A
No symbol = official figure, P = official figure, F = FAO estimate, * = Unofficial/Semi-official/mirror data, C = Calculated figure A = Aggregate (may include official, semi-official or estimates);			
Source: UN Food & Agriculture Organisation (FAO)			

Carrot

The carrot (Daucus carota subsp. sativus) is a root vegetable, usually orange in colour, though purple, black, red, white, and yellow varieties exist. Carrots are a domesticated form of the wild carrot, Daucus carota, native to Europe and southwestern Asia. The plant probably originated in Persia and originally cultivated for its leaves and seeds. The most commonly eaten part of the plant is the taproot, although the greens are sometimes eaten as well. The domestic carrot has been selectively bred for its greatly enlarged, more palatable, less woody-textured taproot.

The carrot is a biennial plant in the umbellifer family Apiaceae. At first, it grows a rosette of leaves while building up the enlarged taproot. Fast-growing varieties mature within three months (90 days) of sowing the seed, while slower-maturing varieties are harvested four months later (120 days). The roots contain high quantities of alpha- and beta-carotene, and are a good source of vitamin K and vitamin B6, but the belief that eating carrots improves night vision is a myth put forward by the British in World War II to mislead the enemy about their military capabilities.

The United Nations Food and Agriculture Organization (FAO) reports that world production of carrots and turnips (these plants are combined by the FAO) for the calendar year 2013 was 37.2 million tonnes; almost half (~45%) were grown in China. Carrots are widely used in many cuisines, especially in the preparation of salads, and carrot salads are a tradition in many regional cuisines.

History

Molecular and genetic studies, along with written history, support the idea that the cultivated/domesticated carrot has a single origin in Central Asia. The wild ancestors of the carrot are likely to have originated in Persia (regions of which are now Iran and

Afghanistan), which remains the centre of diversity for Daucus carota, the wild carrot. A naturally occurring subspecies of the wild carrot was presumably bred selectively over the centuries to reduce bitterness, increase sweetness and minimise the woody core; this process produced the familiar garden vegetable.

A depiction labeled "garden" carrot from the Juliana Anicia Codex, a copy, written in Constantinople in 515 AD, of Dioscorides' 1st century AD Greek work. The facing page states that "the root can be cooked and eaten."

When they were first cultivated, carrots were grown for their aromatic leaves and seeds rather than their roots. Carrot seeds have been found in Switzerland and Southern Germany dating back to 2000–3000 BC. Some close relatives of the carrot are still grown for their leaves and seeds, for example: parsley, cilantro/coriander, fennel, dill and cumin. The first mention of the root in classical sources is during the 1st century, and the carrot may have been eaten as a root vegetable by the Romans, although there is some ambiguity about this, as they used the word pastinaca for both carrots and parsnips, part of the same family.

The plant is depicted and described in the Eastern Roman Juliana Anicia Codex, a copy, made in 515 AD in Constantinople, of the Greek physician Dioscorides' 1st century AD pharmacopoeia of herbs and medicines, De Materia Medica. Three different types of carrots are depicted, and the text states of them that "the root can be cooked and eaten."

The plant appears to have been introduced into Spain by the Moors in the 8th century. In the 10th century, in wordwide locations like West Asia, India and Europe, the roots were purple. The modern carrot originated in Afghanistan at about this time. The Jewish scholar Simeon Seth describes both red and yellow carrots in the 11th century. The 12th-century Arab-Andalusian agriculturist, Ibn al-'Awwam, also mentions roots of these colours; Cultivated carrots appeared in China in the 14th century, and in Japan in the 18th century. Orange-coloured carrots appeared in the Netherlands in the 17th century, which has been related to the fact that the Dutch flag at the time, the Prince's Flag, included orange. These, the modern carrots, were intended by the English antiquary John Aubrey (1626–1697) when he noted in his memoranda, "Carrots were first sown at Beckington in Somersetshire. Some very old Man there [in 1668] did remember their first bringing hither." European settlers introduced the carrot to colonial America in the 17th century.

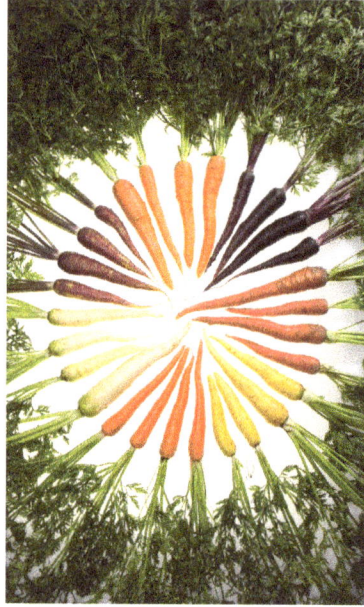

Carrots in a range of colours

Outwardly purple carrots, still orange on the inside, were sold in British stores beginning in 2002.

Description

Daucus carota is a biennial plant that grows a rosette of leaves while building up the taproot that stores large amounts of sugars to provide energy for the plant to flower in the second year.

Seedlings shortly after germination

Soon after germination, carrot seedlings show a distinct demarcation between taproot and stem: the stem is thicker and lacks lateral roots. At the upper end of the stem is the seed leaf. The first true leaf appears about 10–15 days after germination. Subsequent leaves, produced from the stem nodes, are alternating (with a single leaf attached to a node) and compound, and arranged in a spiral. The alternating compound leaves show pinnation. As the plant grows, the bases of the seed leaves, near the taproot, are pushed apart. The stem, located just above the ground, is compressed and the internodes are not distinct. When the seed stalk elongates for flowering, the tip of the stem narrows and becomes pointed, extends upward, and becomes a highly branched inflorescence. The tall stems grow to 60–200 cm (20–80 in) tall.

Most of the taproot consists of a pulpy outer cortex (phloem) and an inner core (xylem). High-quality carrots have a large proportion of cortex compared to core. Although a completely xylem-free carrot is not possible, some cultivars have small and deeply pigmented cores; the taproot can appear to lack a core when the colour of the cortex and core are similar in intensity. Taproots typically have a long conical shape, although cylindrical and round cultivars are available. The root diameter can range from 1 cm (0.4 in) to as much as 10 cm (4 in) at the widest part. The root length ranges from 5 to 50 cm (2.0 to 19.7 in), although most are between 10 and 25 cm (4 and 10 in).

Daucus carota umbel (inflorescence). Individual flowers are borne on undivided pedicels originating from a common node.

Top view of Daucus carota inflorescence, showing umbellets; the central flower is dark red

Flower development begins when the flat meristem changes from producing leaves to an uplifted, conical meristem capable of producing stem elongation and a cluster of flowers. The cluster is a compound umbel, and each umbel contains several smaller umbels (umbellets). The first (primary) umbel occurs at the end of the main floral stem; smaller secondary umbels grow from the main branch, and these further branch into third, fourth, and even later-flowering umbels. A large, primary umbel can contain up to 50 umbellets, each of which may have as many as 50 flowers; subsequent umbels have fewer flowers. Individual flowers are small and white, sometimes with a light green or yellow tint. They consist of five petals, five stamens, and an entire calyx. The stamens usually split and fall off before the stigma becomes receptive to receive pollen. The stamens of the brown, male, sterile flowers degenerate and shrivel before the flower fully opens. In the other type of male sterile flower, the stamens are replaced by petals, and these petals do not fall off. A nectar-containing disc is present on the upper surface of the carpels.

Flowers consist of five petals, five stamens, and an entire calyx.

Flowers change sex in their development, so the stamens release their pollen before the stigma of the same flower is receptive. The arrangement is centripetal, meaning the oldest flowers are near the edge and the youngest flowers are in the center. Flowers usually first open at the outer edge of the primary umbel, followed about a week later on the secondary umbels, and then in subsequent weeks in higher-order umbels. The usual flowering period of individual umbels is 7 to 10 days, so a plant can be in the process of flowering for 30–50 days. The distinctive umbels and floral nectaries attract pollinating insects. After fertilization and as seeds develop, the outer umbellets of an umbel bend inward causing the umbel shape to change from slightly convex or fairly flat to concave, and when cupped it resembles a bird's nest.

The fruit that develops is a schizocarp consisting of two mericarps; each mericarp is a true seed. The paired mericarps are easily separated when they are dry. Premature separation (shattering) before harvest is undesirable because it can result in seed loss. Mature seeds are flattened on the commissural side that faced the septum of the ovary. The flattened side has five longitudinal ribs. The bristly hairs that protrude from some ribs are usually removed by abrasion during milling and cleaning. Seeds also contain oil ducts and canals. Seeds vary somewhat in size, ranging from less than 500 to more than 1000 seeds per gram.

The carrot is a diploid species, and has nine relatively short, uniform-length chromosomes (2n=18). The genome size is estimated to be 473 mega base pairs, which is four times larger than Arabidopsis thaliana, one-fifth the size of the maize genome, and about the same size as the rice genome.

Chemistry

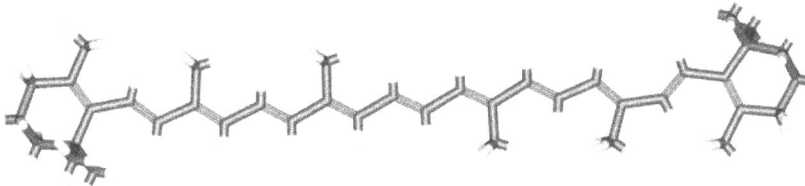

β-Carotene structure. Carotene is responsible for the orange colour of carrots and many other fruits and vegetables.

Polyacetylenes can be found in Apiaceae vegetables like carrots where they show cytotoxic activities. Falcarinol and falcarindiol (cis-heptadeca-1,9-diene-4,6-diyne-3,8-diol) are such compounds. This latter compound shows antifungal activity towards Mycocentrospora acerina and Cladosporium cladosporioides. Falcarindiol is the main compound responsible for bitterness in carrots.

Other compounds such as pyrrolidine (present in the leaves), 6-hydroxymellein, 6-methoxymellein, eugenin, 2,4,5-trimethoxybenzaldehyde (gazarin) or (Z)-3-acetoxy-heptadeca-1,9-diene-4,6-diin-8-ol (falcarindiol 3-acetate) can also be found in carrot.

Cultivation

Carrots are grown from seed and can take up to four months (120 days) to mature, but most varieties mature within 70 to 80 days under the right conditions. They grow best in full sun but tolerate some shade. The optimum temperature is 16 to 21 °C (61 to 70 °F). The ideal soil is deep, loose and well-drained, sandy or loamy, with a pH of 6.3 to 6.8. Fertilizer should be applied according to soil type because the crop requires low levels of nitrogen, moderate phosphate and high potash. Rich or rocky soils should be avoided, as these will cause the roots to become hairy and/or misshapen. Irrigation is applied when needed to keep the soil moist. After sprouting, the crop is eventually thinned to a spacing of 8 to 10 cm (3 to 4 in) and weeded to prevent competition beneath the soil.

Workers harvesting carrots, Imperial Valley, California, 1948

Cultivation Problems

There are several diseases that can reduce the yield and market value of carrots. The most devastating carrot disease is Alternaria leaf blight, which has been known to eradicate entire crops. A bacterial leaf blight caused by Xanthomonas campestris can also be destructive in warm, humid areas. Root knot nematodes (Meloidogyne species) can cause stubby or forked roots, or galls. Cavity spot, caused by the oomycetes Pythium violae and Pythium sulcatum, results in irregularly shaped, depressed lesions on the taproots.

Physical damage can also reduce the value of carrot crops. The two main forms of damage are splitting, whereby a longitudinal crack develops during growth that can be a few centimetres to the entire length of the root, and breaking, which occurs postharvest. These disorders can affect over 30% of commercial crops. Factors associated with high levels of splitting include wide plant spacing, early sowing, lengthy growth durations, and genotype.

Companion Planting

Carrots benefit from strongly scented companion plants. The pungent odour of onions, leeks and chives help repel the carrot root fly, and other vegetables that team well with carrots include lettuce, tomatoes and radishes, as well as the herbs rosemary and sage. Carrots thrive in the presence of caraway, coriander, chamomile, marigold and Swan River daisy. They can also be good companions for other plants; if left to flower, the carrot, like any umbellifer, attracts predatory wasps that kill many garden pests.

Cultivars

Carrot seeds

Seeds of Daucus carota subsp. maximus - MHNT

Carrot cultivars can be grouped into two broad classes, eastern carrots and western carrots. A number of novelty cultivars have been bred for particular characteristics.

"Eastern" (a European and American continent reference) carrots were domesticated in Persia (probably in the lands of modern-day Iran and Afghanistan within West Asia) during the 10th century, or possibly earlier. Specimens of the "eastern" carrot that survive to the present day are commonly purple or yellow, and often have branched roots. The purple colour common in these carrots comes from anthocyanin pigments.

The western carrot emerged in the Netherlands in the 17th century, There is a popular belief that its orange colour making it popular in those countries as an emblem of the House of Orange and the struggle for Dutch independence, although there is little evidence for this. The orange colour results from abundant carotenes in these cultivars.

Western carrot cultivars are commonly classified by their root shape. The four general types are:

- Chantenay carrots. Although the roots are shorter than other cultivars, they have vigorous foliage and greater girth, being broad in the shoulders and tapering towards a blunt, rounded tip. They store well, have a pale-coloured core and are mostly used for processing. Varieties include Carson Hybrid and Red Cored Chantenay.

- Danvers carrots. These have strong foliage and the roots are longer than Chantaney types, and they have a conical shape with a well-defined shoulder, tapering to a point. They are somewhat shorter than Imperator cultivars, but more tolerant of heavy soil conditions. Danvers cultivars store well and are used both fresh and for processing. They were developed in 1871 in Danvers, Massachusetts. Varieties include Danvers Half Long and Danvers 126.

- Imperator carrots. This cultivar has vigorous foliage, is of high sugar content, and has long and slender roots, tapering to a pointed tip. Imperator types are the most widely cultivated by commercial growers. Varieties include Imperator 58 and Sugarsnax Hybrid.

- Nantes carrots. These have sparse foliage, are cylindrical, short with a more blunt tip than Imperator types, and attain high yields in a range of conditions. The skin is easily damaged and the core is deeply pigmented. They are brittle, high in sugar and store less well than other types. Varieties include Nelson Hybrid, Scarlet Nantes and Sweetness Hybrid.

One particular variety lacks the usual orange pigment due to carotene, and owing its white colour to a recessive gene for tocopherol (vitamin E). Derived from Daucus carota L. and patented at the University of Wisconsin–Madison, the variety is intended to supplement the dietary intake of Vitamin E.

Production

Production of carrots and turnips – 2013	
Country	Production (millions of tonnes)
China	16.8
Uzbekistan	1.6
Russia	1.6
United States	1.3
Ukraine	0.9
World	37.2
Source: FAOSTAT of the United Nations	

Carrots are one of the ten most economically important vegetable crops in the world. In 2013, world production of carrots (combined with turnips) was 37.2 million tonnes, with China producing 45% of the world total (16.8 million tonnes, table). Other major producers were Uzbekistan and Russia (4% of world total each), the United States (3%) and Ukraine (2%).

Storage

Carrots can be stored for several months in the refrigerator or over winter in a moist, cool place. For long term storage, unwashed carrots can be placed in a bucket between layers of sand, a 50/50 mix of sand and wood shavings, or in soil. A temperature range of 32 to 40 °F (0 to 5 °C) is best.

Consumption

Carrot tzimmes

Carrots can be eaten in a variety of ways. Only 3 percent of the β-carotene in raw carrots is released during digestion: this can be improved to 39% by pulping, cooking and adding cooking oil. Alternatively they may be chopped and boiled, fried or steamed, and cooked in soups and stews, as well as baby and pet foods. A well-known dish is carrots julienne. Together with onion and celery, carrots are one of the primary vegetables used in a mirepoix to make various broths.

The greens are edible as a leaf vegetable, but are rarely eaten by humans; some sources suggest that the greens contain toxic alkaloids. When used for this purpose, they are harvested young in high-density plantings, before significant root development, and typically used stir-fried, or in salads. Some people are allergic to carrots. In a 2010 study on the prevalence of food allergies in Europe, 3.6 percent of young adults showed some degree of sensitivity to carrots. Because the major carrot allergen, the protein Dauc c 1.0104, is cross-reactive with homologues in birch pollen (Bet v 1) and mugwort pollen (Art v 1), most carrot allergy sufferers are also allergic to pollen from these plants.

In India carrots are used in a variety of ways, as salads or as vegetables added to spicy rice or dal dishes. A popular variation in north India is the Gajar Ka Halwa carrot dessert, which has carrots grated and cooked in milk until the whole mixture is solid, after which nuts and butter are added. Carrot salads are usually made with grated carrots with a seasoning of mustard seeds and green chillies popped in hot oil. Carrots can also be cut in thin strips and added to rice, can form part of a dish of mixed roast vegetables or can be blended with tamarind to make chutney.

Since the late 1980s, baby carrots or mini-carrots (carrots that have been peeled and cut into uniform cylinders) have been a popular ready-to-eat snack food available in many supermarkets. Carrots are puréed and used as baby food, dehydrated to make chips, flakes, and powder, and thinly sliced and deep-fried, like potato chips.

The sweetness of carrots allows the vegetable to be used in some fruit-like roles. Grated carrots are used in carrot cakes, as well as carrot puddings, an English dish thought to have originated in the early 19th century. Carrots can also be used alone or blended with fruits in jams and preserves. Carrot juice is also widely marketed, especially as a health drink, either stand-alone or blended with juices extracted from fruits and other vegetables.

Highly excessive consumption over a period of time results in a condition of carotenemia which is a yellowing of the skin caused by a build up of carotenoids.

Nutrition

Carrots, raw	
Nutritional value per 100 g (3.5 oz)	
Energy	173 kJ (41 kcal)
Carbohydrates	9.6 g
Sugars	4.7 g
Dietary fiber	2.8 g
Fat	0.24 g
Protein	0.93 g
Vitamins	
Vitamin A equiv.	(104%)
	835 µg
beta-carotene	(77%)
lutein zeaxanthin	8285 µg
	256 µg
Thiamine (B1)	(6%)
	0.066 mg

Riboflavin (B2)	(5%)	0.058 mg
Niacin (B3)	(7%)	0.983 mg
Pantothenic acid (B5)	(5%)	0.273 mg
Vitamin B6	(11%)	0.138 mg
Folate (B9)	(5%)	19 µg
Vitamin C	(7%)	5.9 mg
Vitamin E	(4%)	0.66 mg
Vitamin K	(13%)	13.2 µg
Minerals		
Calcium	(3%)	33 mg
Iron	(2%)	0.3 mg
Magnesium	(3%)	12 mg
Manganese	(7%)	0.143 mg
Phosphorus	(5%)	35 mg
Potassium	(7%)	320 mg

Sodium	(5%)	
	69 mg	
Zinc	(3%)	
	0.24 mg	
Other constituents		
Fluoride	3.2 µg	
Water	88 g	
Link to USDA Database Entry		
• Units		
• µg = micrograms • mg = milligrams		
• IU = International units		
Percentages are roughly approximated using US recommendations for adults.		

The carrot gets its characteristic, bright orange colour from β-carotene, and lesser amounts of α-carotene, γ-carotene, lutein and zeaxanthin. α- and β-carotenes are partly metabolized into vitamin A, providing more than 100% of the Daily Value (DV) per 100 g serving of carrots (right table). Carrots are also a good source of vitamin K (13% DV) and vitamin B6 (11% DV), but otherwise have modest content of other essential nutrients (right table).

Carrots are 88% water, 4.7% sugar, 0.9% protein, 2.8% dietary fiber, 1% ash and 0.2% fat. Carrot dietary fiber comprises mostly cellulose, with smaller proportions of hemicellulose, lignin and starch. Free sugars in carrot include sucrose, glucose and fructose.

The lutein and zeaxanthin carotenoids characteristic of carrots are studied for their potential roles in vision and eye health.

Night Vision

The provitamin A beta-carotene from carrots does not actually help people to see in the dark unless they suffer from a deficiency of vitamin A. This myth was propaganda used by the Royal Air Force during the Second World War to explain why their pilots had improved success during night air battles, but was actually used to disguise advances in radar technology and the use of red lights on instrument panels. Nevertheless, the consumption of carrots was advocated in Britain at the time as part of a Dig for Victory campaign. A radio programme called The Kitchen Front encouraged people to grow, store and use carrots in various novel ways, including making carrot jam and Woolton pie, named after the Lord Woolton, the Minister for Food.

References

- Graeme Barker (25 March 2009). The Agricultural Revolution in Prehistory: Why did Foragers become Farmers?. Oxford University Press. ISBN 978-0-19-955995-4. Retrieved 15 August 2012.

- Miller, Richard (1985). The Potential of Herbs as a Cash Crop. Kansas City:MO: Acres U.S.A. p. 6. ISBN 0911311106.

- Traunfeld, Jerry (2000). "The Herbfarm cookbook". New York: Scribner. ISBN 0-684-83976-8. Retrieved January 15, 2012.

- Miller, Richard Alan (2000). "Getting started : important considerations for the herb farmer". Goodwood, Ont.: Richters. Retrieved January 15, 2012. ISBN 1-894021-04-5

- Bown, Deni (1995). Encyclopedia of Herbs and Their Uses. New York: Dorling Kindersley. p. 12. ISBN 0789401843.

- Gardner, Jo Ann (1997). Living with Herbs: a treasury of useful plants for the home and garden. Woodstock:VT: The Countryman Press. p. 30. ISBN 0881503592.

- Shores, Sandie (1999). Growing and Selling Fresh-cut Herbs. Pownal:VT: Storey Books. p. 78. ISBN 1580171281.

- Aliza Green (January 2006). Field Guide to Herbs & Spices: How to Identify, Select, and Use Virtually Every Seasoning at the Market. Quirk Books. pp. 116–117. ISBN 978-1-59474-082-4.

- Aggarwal, Bharat B.; Kunnumakkara, Ajaikumar B. (2009). Molecular Targets and Therapeutic Uses of Spices: Modern Uses for Ancient Medicine. Singapore: World Scientific Publishing. p. 150. ISBN 978-981-283-790-5.

- Samuelsson, Marcus (2003). Aquavit: And the New Scandinavian Cuisine. Houghton Mifflin Harcourt. p. 12 (of 312). ISBN 0618109412.

- Crawford, Martin 2010. Creating a Forest Garden: Working with Nature to Grow Edible Crops. Totnes: Green Books. ISBN 1-900322-62-5.

- Douglas, J. Sholto and Hart, Robert A. de J. 1985. Forest Farming. Intermediate Technology. ISBN 0-946688-30-3.

3

Orchard: A Comprehensive Study

An orchard is the planting of trees for food production. They comprise of fruit producing trees or nut producing trees; they also feature large gardens that serve an aesthetic purpose. The chapter will not only provide an outline, it will also delve deep into the topics related to it.

Orchard

A lemon orchard in the Upper Galilee in Israel.

An orchard is an intentional planting of trees or shrubs that is maintained for food production. Orchards comprise fruit- or nut-producing trees which are generally grown for commercial production. Orchards are also sometimes a feature of large gardens, where they serve an aesthetic as well as a productive purpose. A fruit garden is generally synonymous with an orchard, although it is set on a smaller non-commercial scale and may emphasize berry shrubs in preference to fruit trees. Most temperate-zone orchards are laid out in a regular grid, with a grazed or mown grass or bare soil base that makes maintenance and fruit gathering easy.

Orchards are sometimes concentrated near bodies of water, where climatic extremes are moderated and blossom time is retarded until frost danger is past.

Layout

An orchard's layout is the technique of planting the crops in a proper system. There are different methods of planting and thus different layouts. Some of these layout types include:

1. Square method

2. Rectangular method

3. Quincunx method

4. Triangular method

5. Hexagonal method

6. Contour (or Terracing) method

For different varieties, these systems may vary to some extent.

Orchards by Region

Apple orchards in Azwell, Washington surrounding a community of pickers' cabins

Sour cherry orchard on Lake Erie shoreline (Leamington, Ontario)

The most extensive orchards in the United States are apple and orange orchards, although citrus orchards are more commonly called groves. The most extensive apple orchard area is in eastern Washington state, with a lesser but significant apple orchard area in most of Upstate New York. Extensive orange orchards are found in Florida and southern California,where they are more widely known as 'groves'. In eastern North America, many orchards are along the shores of Lake Michigan (such as the Fruit Ridge Region), Lake Erie, and Lake Ontario.

In Canada, apple and other fruit orchards are widespread on the Niagara Peninsula, south of Lake Ontario. This region is known as Canada Fruitbelt and, in addition to large-scale commercial fruit marketing, it encourages "pick-your-own" activities in the harvest season.

Murcia is a major orchard area (or la huerta) in Europe, with citrus crops. New Zealand, China, Argentina and Chile also have extensive apple orchards.

Tenbury Wells in Worcestershire has been called The Town in the Orchard, since the 19th century, because it was surrounded by extensive orchards. Today, this heritage is celebrated through an annual Applefest.

Central Europe

Streuobstwiese (pl. Streuobstwiesen) is a German word that means a meadow with scattered fruit trees or fruit trees that are planted in a field. Streuobstwiese, or a meadow orchard, is a traditional landscape in the temperate, maritime climate of continental Western Europe. In the 19th and early 20th centuries, Streuobstwiesen were a kind of a rural community orchard that were intended for productive cultivation of stone fruit. In recent years, ecologists have successfully lobbied for state subsidies to valuable habitats, biodiversity and natural landscapes, which are also used to preserve old meadow orchards. Both conventional and meadow orchards provide a suitable habitat for many animal species that live in a cultured landscape. A notable example is the hoopoe that nests in tree hollows of old fruit trees and, in the absence of alternative nesting sites, is threatened in many parts of Europe, because of the destruction of old orchards.

Historical Orchards

Old growth apple orchard in Ottawa, Canada

- Orchard House in Concord, Massachusetts was the residence of American celebrated writer Louisa May Alcott.

- Fruita, Utah part of Capitol Reef National Park has Mormon pioneer orchards maintained by the United States National Park Service.

Orchard Conservation in the UK

- Natural England, through its Countryside Stewardship Scheme, Environmental

Stewardship and Environmentally Sensitive Areas Scheme, gives grant aid and advice for the maintenance, enhancement or re-creation of historical orchards.

- The 'Orchard Link' organisation provides advice on how to manage and restore the county of Devon's orchards, as well as enabling the local community to use the local orchard produce. An organisation called 'Orchards Live' carries out similar work in North Devon.

- People's Trust for Endangered Species (PTES) has mapped every traditional orchard within England and Wales and manages the national inventory for this habitat.

- The UK Biodiversity Partnership lists traditional orchards and a priority UK Biodiversity Action Plan habitat.

- The Wiltshire Traditional Orchards Project maps, conserves and restores traditional orchards within Wiltshire, England.

Fruit

Culinary fruits

Several culinary fruits

In botany, a fruit is the seed-bearing structure in flowering plants (also known as angiosperms) formed from the ovary after flowering.

Mixed fruit

The Medici citrus collection by Bartolomeo Bimbi, 1715

Fruits are the means by which angiosperms disseminate seeds. Edible fruits, in particular, have propagated with the movements of humans and animals in a symbiotic relationship as a means for seed dispersal and nutrition; in fact, humans and many animals have become dependent on fruits as a source of food. Accordingly, fruits account for a substantial fraction of the world's agricultural output, and some (such as the apple and the pomegranate) have acquired extensive cultural and symbolic meanings.

In common language usage, "fruit" normally means the fleshy seed-associated structures of a plant that are sweet or sour, and edible in the raw state, such as apples, bananas, grapes, lemons, oranges, and strawberries. On the other hand, in botanical usage, "fruit" includes many structures that are not commonly called "fruits", such as bean pods, corn kernels, tomatoes, and wheat grains. The section of a fungus that produces spores is also called a fruiting body.

Botanic Fruit and Culinary Fruit

Many common terms for seeds and fruit do not correspond to the botanical classifications. In culinary terminology, a fruit is usually any sweet-tasting plant part, especially a botanical fruit; a nut is any hard, oily, and shelled plant product; and a vegetable is any savory or less sweet plant product. However, in botany, a fruit is the ripened ovary or carpel that contains seeds, a nut is a type of fruit and not a seed, and a seed is a ripened ovule.

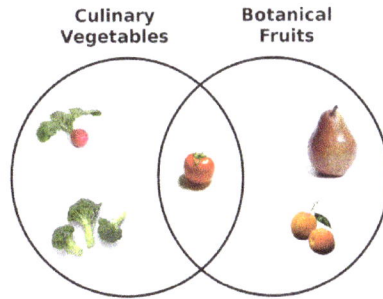

Venn diagram representing the relationship between (culinary) vegetables and botanical fruits

Examples of culinary "vegetables" and nuts that are botanically fruit include corn, cucurbits (e.g., cucumber, pumpkin, and squash), eggplant, legumes (beans, peanuts, and peas), sweet pepper, and tomato. In addition, some spices, such as allspice and chili pepper, are fruits, botanically speaking. In contrast, rhubarb is often referred to as a fruit, because it is used to make sweet desserts such as pies, though only the petiole (leaf stalk) of the rhubarb plant is edible, and edible gymnosperm seeds are often given fruit names, e.g., ginkgo nuts and pine nuts.

Botanically, a cereal grain, such as corn, rice, or wheat, is also a kind of fruit, termed a caryopsis. However, the fruit wall is very thin and is fused to the seed coat, so almost all of the edible grain is actually a seed.

Fruit Structure

The outer, often edible layer, is the pericarp, formed from the ovary and surrounding the seeds, although in some species other tissues contribute to or form the edible portion. The pericarp may be described in three layers from outer to inner, the epicarp, mesocarp and endocarp.

Fruit that bears a prominent pointed terminal projection is said to be beaked.

Fruit Development

A fruit results from maturation of one or more flowers, and the gynoecium of the flower(s) forms all or part of the fruit.

Inside the ovary/ovaries are one or more ovules where the megagametophyte contains the egg cell. After double fertilization, these ovules will become seeds. The ovules are fertilized in a process that starts with pollination, which involves the movement of pollen from the stamens to the stigma of flowers. After pollination, a tube grows from the pollen through the stigma into the ovary to the ovule and two sperm are transferred from the pollen to the megagametophyte. Within the megagametophyte one of the two sperm unites with the egg, forming a zygote, and the second sperm enters the central cell forming the endosperm mother cell, which completes the double fertilization process. Later the zygote will give rise to the embryo of the seed, and the endosperm mother cell will give rise to endosperm, a nutritive tissue used by the embryo.

The development sequence of a typical drupe, the nectarine (Prunus persica) over a 7.5 month period, from bud formation in early winter to fruit ripening in midsummer

As the ovules develop into seeds, the ovary begins to ripen and the ovary wall, the pericarp, may become fleshy (as in berries or drupes), or form a hard outer covering (as in nuts). In some multiseeded fruits, the extent to which the flesh develops is proportional to the number of fertilized ovules. The pericarp is often differentiated into two or three distinct layers called the exocarp (outer layer, also called epicarp), mesocarp (middle layer), and endocarp (inner layer). In some fruits, especially simple fruits derived from an inferior ovary, other parts of the flower (such as the floral tube, including the petals, sepals, and stamens), fuse with the ovary and ripen with it. In other cases, the sepals, petals and/or stamens and style of the flower fall off. When such other floral parts are a significant part of the fruit, it is called an accessory fruit. Since other parts of the flower may contribute to the structure of the fruit, it is important to study flower structure to understand how a particular fruit forms.

There are three general modes of fruit development:

- Apocarpous fruits develop from a single flower having one or more separate carpels, and they are the simplest fruits.

- Syncarpous fruits develop from a single gynoecium having two or more carpels fused together.

- Multiple fruits form from many different flowers.

Plant scientists have grouped fruits into three main groups, simple fruits, aggregate fruits, and composite or multiple fruits. The groupings are not evolutionarily relevant, since many diverse plant taxa may be in the same group, but reflect how the flower organs are arranged and how the fruits develop.

Simple Fruit

Epigynous berries are simple fleshy fruit. Clockwise from top right: cranberries, lingonberries, blueberries, red huckleberries

Simple fruits can be either dry or fleshy, and result from the ripening of a simple or compound ovary in a flower with only one pistil. Dry fruits may be either dehiscent (they open to discharge seeds), or indehiscent (they do not open to discharge seeds). Types of dry, simple fruits, and examples of each, include:

- achene – Most commonly seen in aggregate fruits (e.g., strawberry)

- capsule – (e.g., Brazil nut)

- caryopsis – (e.g., wheat)

- cypsela – an achene-like fruit derived from the individual florets in a capitulum (e.g., dandelion).

- fibrous drupe – (e.g., coconut, walnut)

- follicle – is formed from a single carpel, opens by one suture (e.g., milkweed), commonly seen in aggregate fruits (e.g., magnolia)

- legume – (e.g., bean, pea, peanut)

- loment – a type of indehiscent legume

- nut – (e.g., beech, hazelnut, oak acorn)

- samara – (e.g., ash, elm, maple key)

- schizocarp – (e.g., carrot seed)

- silique – (e.g., radish seed)

- silicle – (e.g., shepherd's purse)

- utricle – (e.g., beet)

Lilium unripe capsule fruit

Fruits in which part or all of the pericarp (fruit wall) is fleshy at maturity are simple fleshy fruits. Types of simple, fleshy, fruits (with examples) include:

- berry – (e.g., cranberry, gooseberry, redcurrant, tomato)

- stone fruit or drupe (e.g., apricot, cherry, olive, peach, plum)

Dewberry flowers. Note the multiple pistils, each of which will produce a drupelet. Each flower will become a blackberry-like aggregate fruit.

An aggregate fruit, or etaerio, develops from a single flower with numerous simple pistils.

- Magnolia and peony, collection of follicles developing from one flower.

- Sweet gum, collection of capsules.

- Sycamore, collection of achenes.

- Teasel, collection of cypsellas

- Tuliptree, collection of samaras.

The pome fruits of the family Rosaceae, (including apples, pears, rosehips, and saskatoon berry) are a syncarpous fleshy fruit, a simple fruit, developing from a half-inferior ovary.

Schizocarp fruits form from a syncarpous ovary and do not really dehisce, but rather split into segments with one or more seeds; they include a number of different forms from a wide range of families. Carrot seed is an example.

Aggregate Fruit

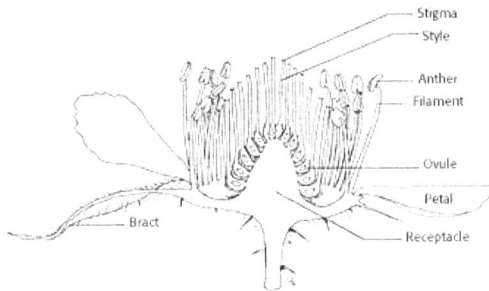

Figure 167. - Longitudinal section of 'Willamett' raspberry flower, x10.

Detail of raspberry flower

Aggregate fruits form from single flowers that have multiple carpels which are not joined together, i.e. each pistil contains one carpel. Each pistil forms a fruitlet, and collectively the fruitlets are called an etaerio. Four types of aggregate fruits include etaerios of achenes, follicles, drupelets, and berries. Ranunculaceae species, including Clematis and Ranunculus have an etaerio of achenes, Calotropis has an etaerio of follicles, and Rubus species like raspberry, have an etaerio of drupelets. Annona have an etaerio of berries.

The raspberry, whose pistils are termed drupelets because each is like a small drupe attached to the receptacle. In some bramble fruits (such as blackberry) the receptacle is elongated and part of the ripe fruit, making the blackberry an aggregate-accessory fruit. The strawberry is also an aggregate-accessory fruit, only one in which the seeds are contained in achenes. In all these examples, the fruit develops from a single flower with numerous pistils.

Multiple Fruits

A multiple fruit is one formed from a cluster of flowers (called an inflorescence). Each flower produces a fruit, but these mature into a single mass. Examples are the pineapple, fig, mulberry, osage-orange, and breadfruit.

In the photograph on the right, stages of flowering and fruit development in the noni or Indian mulberry (Morinda citrifolia) can be observed on a single branch. First an inflorescence of white flowers called a head is produced. After fertilization, each flower develops into a drupe, and as the drupes expand, they become connate (merge) into a multiple fleshy fruit called a syncarp.

In some plants, such as this noni, flowers are produced regularly along the stem and it is possible to see together examples of flowering, fruit development, and fruit ripening.

Berries

Berries are another type of fleshy fruit; they are simple fruit created from a single ovary. The ovary may be compound, with several carpels. Types include (examples follow in the table below):

- Pepo – berries whose skin is hardened, cucurbits

- Hesperidium – berries with a rind and a juicy interior, like most citrus fruit

Accessory Fruit

The fruit of a pineapple includes tissue from the sepals as well as the pistils of many flowers. It is an accessory fruit and a multiple fruit.

Some or all of the edible part of accessory fruit is not generated by the ovary. Accessory fruit can be simple, aggregate, or multiple, i.e., they can include one or more pistils and other parts from the same flower, or the pistils and other parts of many flowers.

Table of fruit examples

Types of fleshy fruits					
True berry	Pepo	Hesperidium	Aggregate fruit	Multiple fruit	Accessory fruit
Blackcurrant, Blueberry, Chili pepper, Cranberry, Eggplant, Gooseberry, Grape, Guava, Kiwifruit, Lucuma, Pomegranate, Redcurrant, Tomato	Cucumber, Gourd, Melon, Pumpkin	Grapefruit, Lemon, Lime, Orange	Blackberry, Boysenberry, Raspberry	Fig, Hedge apple, Mulberry, Pineapple	Apple, Pineapple, Rose hip, Stone fruit, Strawberry

Seedless Fruits

An arrangement of fruits commonly thought of as vegetables, including tomatoes and various squash

Seedlessness is an important feature of some fruits of commerce. Commercial cultivars of bananas and pineapples are examples of seedless fruits. Some cultivars of citrus fruits (especially grapefruit, mandarin oranges, navel oranges), satsumas, table grapes, and watermelons are valued for their seedlessness. In some species, seedlessness is the result of parthenocarpy, where fruits set without fertilization. Parthenocarpic fruit set may or may not require pollination, but most seedless citrus fruits require a stimulus from pollination to produce fruit.

Seedless bananas and grapes are triploids, and seedlessness results from the abortion of the embryonic plant that is produced by fertilization, a phenomenon known as stenospermocarpy, which requires normal pollination and fertilization.

Seed Dissemination

Variations in fruit structures largely depend on their seeds' mode of dispersal. This dispersal can be achieved by animals, explosive dehiscence, water, or wind.

Some fruits have coats covered with spikes or hooked burrs, either to prevent themselves from being eaten by animals, or to stick to the feathers, hairs, or legs of animals, using them as dispersal agents. Examples include cocklebur and unicorn plant.

Grapes and Mangoes

The sweet flesh of many fruits is "deliberately" appealing to animals, so that the seeds held within are eaten and "unwittingly" carried away and deposited (i.e., defecated) at a distance from the parent. Likewise, the nutritious, oily kernels of nuts are appealing to rodents (such as squirrels), which hoard them in the soil to avoid starving during the winter, thus giving those seeds that remain uneaten the chance to germinate and grow into a new plant away from their parent.

Other fruits are elongated and flattened out naturally, and so become thin, like wings or helicopter blades, e.g., elm, maple, and tuliptree. This is an evolutionary mechanism to increase dispersal distance away from the parent, via wind. Other wind-dispersed fruit have tiny "parachutes", e.g., dandelion, milkweed, salsify.

Coconut fruits can float thousands of miles in the ocean to spread seeds. Some other fruits that can disperse via water are nipa palm and screw pine.

Some fruits fling seeds substantial distances (up to 100 m in sandbox tree) via explosive dehiscence or other mechanisms, e.g., impatiens and squirting cucumber.

Uses

Nectarines are one of many fruits that can be easily stewed.

Many hundreds of fruits, including fleshy fruits (like apple, kiwifruit, mango, peach, pear, and watermelon) are commercially valuable as human food, eaten both fresh and as jams, marmalade and other preserves. Fruits are also used in manufactured foods (e.g., cakes, cookies, ice cream, muffins, or yogurt) or beverages, such as fruit juices (e.g., apple juice, grape juice, or orange juice) or alcoholic beverages (e.g., brandy, fruit beer, or wine), Fruits are also used for gift giving, e.g., in the form of Fruit Baskets and Fruit Bouquets.

Oranges, bananas, pears, apples, and a watermelon

Many "vegetables" in culinary parlance are botanical fruits, including bell pepper, cucumber, eggplant, green bean, okra, pumpkin, squash, tomato, and zucchini. Olive fruit is pressed for olive oil. Spices like allspice, black pepper, paprika, and vanilla are derived from berries.

Nutritional Value

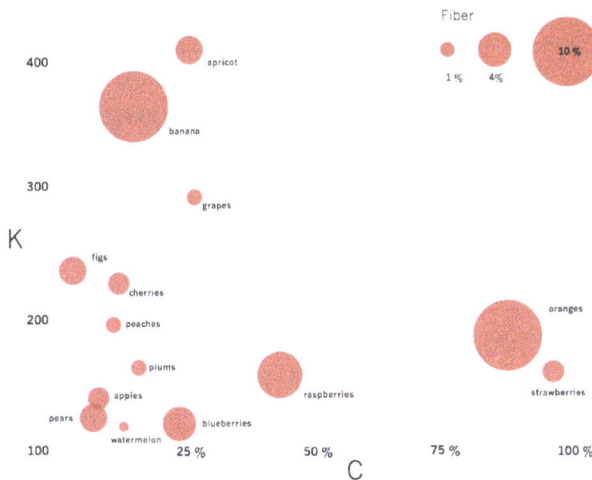

Each point refers to a 100 g serving of the fresh fruit, the daily recommended allowance of vitamin C is on the X axis and mg of Potassium (K) on the Y (offset by 100 mg which every fruit has) and the size of the disk represents amount of fiber (key in upper right). Watermelon, which has almost no fiber, and low levels of vitamin C and potassium, comes in last place.

Fresh fruits are generally high in fiber, vitamin C, and water.

Regular consumption of fruit is generally associated with reduced risks of several diseases and functional declines associated with aging.

Nonfood Uses

Because fruits have been such a major part of the human diet, various cultures have developed many different uses for fruits they do not depend on for food. For example:

- Bayberry fruits provide a wax often used to make candles;

- Many dry fruits are used as decorations or in dried flower arrangements (e.g., annual honesty, cotoneaster, lotus, milkweed, unicorn plant, and wheat). Ornamental trees and shrubs are often cultivated for their colorful fruits, including beautyberry, cotoneaster, holly, pyracantha, skimmia, and viburnum.

- Fruits of opium poppy are the source of opium, which contains the drugs codeine and morphine, as well as the biologically inactive chemical theabaine from which the drug oxycodone is synthesized.

- Osage orange fruits are used to repel cockroaches.

- Many fruits provide natural dyes (e.g., cherry, mulberry, sumac, and walnut).

- Dried gourds are used as bird houses, cups, decorations, dishes, musical instruments, and water jugs.

- Pumpkins are carved into Jack-o'-lanterns for Halloween.

- The spiny fruit of burdock or cocklebur inspired the invention of Velcro.

- Coir fiber from coconut shells is used for brushes, doormats, floor tiles, insulation, mattresses, sacking, and as a growing medium for container plants. The shell of the coconut fruit is used to make bird houses, bowls, cups, musical instruments, and souvenir heads.

- Fruit is often a subject of still life paintings.

Safety

For food safety, the CDC recommends proper fruit handling and preparation to reduce the risk of food contamination and foodborne illness. Fresh fruits and vegetables should be carefully selected; at the store, they should not be damaged or bruised; and pre-cut pieces should be refrigerated or surrounded by ice.

All fruits and vegetables should be rinsed before eating. This recommendation also applies to produce with rinds or skins that are not eaten. It should be done just before preparing or eating to avoid premature spoilage.

Fruits and vegetables should be kept separate from raw foods like meat, poultry, and seafood, as well as from utensils that have come in contact with raw foods. Fruits and vegetables that are not going to be cooked should be thrown away if they have touched raw meat, poultry, seafood, or eggs.

All cut, peeled, or cooked fruits and vegetables should be refrigerated within two hours. After a certain time, harmful bacteria may grow on them and increase the risk of food-borne illness.

Allergies

Fruit allergies make up about 10 percent of all food related allergies

Storage

All fruits benefit from proper post harvest care, and in many fruits, the plant hormone ethylene causes ripening. Therefore, maintaining most fruits in an efficient cold chain is optimal for post harvest storage, with the aim of extending and ensuring shelf life.

Vegetable

Vegetables in a market in the Philippines

In everyday usage, a vegetable is any part of a plant that is consumed by humans as food as part of a savory meal. The term vegetable is somewhat arbitrary, and largely defined through culinary and cultural tradition. It normally excludes other food derived from plants such as fruits, nuts, and cereal grains, but includes seeds such as pulses. The original meaning of the word vegetable, still used in biology, was to describe all types of plant, as in the terms "vegetable kingdom" and "vegetable matter".

Originally, vegetables were collected from the wild by hunter-gatherers and entered cultivation in several parts of the world, probably during the period 10,000 BC to 7,000 BC, when a new agricultural way of life developed. At first, plants which grew locally would have been cultivated, but as time went on, trade brought exotic crops from elsewhere to add to domestic types. Nowadays, most vegetables are grown all over the world as climate permits, and crops may be cultivated in protected environments in less suitable locations. China is the largest producer of vegetables and global trade in agricultural products allows consumers to purchase vegetables grown in faraway countries. The scale of production varies from subsistence farmers supplying the needs of their family for food, to agribusinesses with vast acreages of single-product crops. Depending on the type of vegetable concerned, harvesting the crop is followed by grading, storing, processing, and marketing.

Vegetables can be eaten either raw or cooked and play an important role in human nutrition, being mostly low in fat and carbohydrates, but high in vitamins, minerals and fiber. Many nutritionists encourage people to consume plenty of fruit and vegetables, five or more portions a day often being recommended.

Etymology

Domestic vegetable garden in the United Kingdom

The word vegetable was first recorded in English in the early 15th century. It comes from Old French, and was originally applied to all plants; the word is still used in this sense in biological contexts. It derives from Medieval Latin vegetabilis "growing, flourishing" (i.e. of a plant), a semantic change from a Late Latin meaning "to be enlivening, quickening".

The meaning of "vegetable" as a "plant grown for food" was not established until the 18th century. In 1767, the word was specifically used to mean a "plant cultivated for food, an edible herb or root". The year 1955 noted the first use of the shortened, slang term "veggie".

As an adjective, the word vegetable is used in scientific and technical contexts with a different and much broader meaning, namely of "related to plants" in general, edible or not — as in vegetable matter, vegetable kingdom, vegetable origin, etc.

Terminology

The exact definition of "vegetable" may vary simply because of the many parts of a plant consumed as food worldwide – roots, tubers, bulbs, corms, stems, leaf stems, leaf sheaths, leaves, buds, flowers, fruits, and seeds. The broadest definition is the word's use adjectivally to mean "matter of plant origin" to distinguish it from "animal", meaning "matter of animal origin". More specifically, a vegetable may be defined as "any plant, part of which is used for food", a secondary meaning then being "the edible part of such a plant". A more precise definition is "any plant part consumed for food that is not a fruit or seed, but including mature fruits that are eaten as part of a main meal". Falling outside these definitions are edible fungi (such as mushrooms) and edible seaweed which, although not parts of plants, are often treated as vegetables.

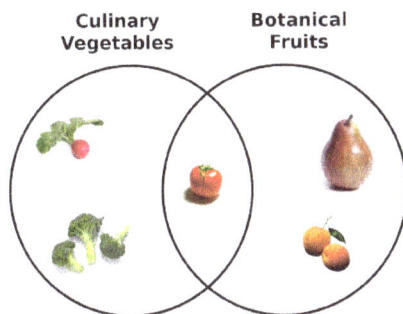

A Venn diagram shows the overlap in the terminology of "vegetables" in a culinary sense and "fruits" in the botanical sense.

In everyday language, the words "fruit" and "vegetable" are mutually exclusive. "Fruit" has a precise botanical meaning, being a part that developed from the ovary of a flowering plant. This is considerably different from the word's culinary meaning. While peaches, plums, and oranges are "fruit" in both senses, many items commonly called "vegetables", such as eggplants, bell peppers, and tomatoes, are botanically fruits. The question of whether the tomato is a fruit or a vegetable found its way into the United States Supreme Court in 1893. The court ruled unanimously in Nix v. Hedden that a tomato is correctly identified as, and thus taxed as, a vegetable, for the purposes of the Tariff of 1883 on imported produce. The court did acknowledge, however, that, botanically speaking, a tomato is a fruit.

History

Before the advent of agriculture, humans were hunter-gatherers. They foraged for edible fruit, nuts, stems, leaves, corms, and tubers, scavenged for dead animals and hunted living ones for food. Forest gardening in a tropical jungle clearing is thought to be the first example of agriculture; useful plant species were identified and encouraged to grow while undesirable species were removed. Plant breeding through the selection of strains with desirable traits such as large fruit and vigorous growth soon followed. While the first evidence for the domestication of grasses such as wheat and barley has been found in the Fertile Crescent in the Middle East, it is likely that various peoples around the world started growing crops in the period 10,000 BC to 7,000 BC. Subsistence agriculture continues to this day, with many rural farmers in Africa, Asia, South

America, and elsewhere using their plots of land to produce enough food for their families, while any surplus produce is used for exchange for other goods.

Throughout recorded history, the rich have been able to afford a varied diet including meat, vegetables and fruit, but for poor people, meat was a luxury and the food they ate was very dull, typically comprising mainly some staple product made from rice, rye, barley, wheat, millet or maize. The addition of vegetable matter provided some variety to the diet. The staple diet of the Aztecs in Central America was maize and they cultivated tomatoes, avocados, beans, peppers, pumpkins, squashes, peanuts, and amaranth seeds to supplement their tortillas and porridge. In Peru, the Incas subsisted on maize in the lowlands and potatoes at higher altitudes. They also used seeds from quinoa, supplementing their diet with peppers, tomatoes, and avocados.

In Ancient China, rice was the staple crop in the south and wheat in the north, the latter made into dumplings, noodles, and pancakes. Vegetables used to accompany these included yams, soybeans, broad beans, turnips, spring onions, and garlic. The diet of the ancient Egyptians was based on bread, often contaminated with sand which wore away their teeth. Meat was a luxury but fish was more plentiful. These were accompanied by a range of vegetables including marrows, broad beans, lentils, onions, leeks, garlic, radishes, and lettuces.

The mainstay of the Ancient Greek diet was bread, and this was accompanied by goat's cheese, olives, figs, fish, and occasionally meat. The vegetables grown included onions, garlic, cabbages, melons, and lentils. In Ancient Rome, a thick porridge was made of emmer wheat or beans, accompanied by green vegetables but little meat, and fish was not esteemed. The Romans grew broad beans, peas, onions and turnips and ate the leaves of beets rather than their roots.

Some Common Vegetables

Some common vegetables					
Image	Description	Parts used	Origin	Cultivars	World production ($\times 10^6$ tons, 2012)
	cabbage Brassica oleracea	leaves, axillary buds, stems, flowerheads	Europe	cabbage, red cabbage, Savoy cabbage, kale, Brussels sprouts, kohlrabi, cauliflower, broccoli, Chinese broccoli	70.1

	turnip Brassica rapa	tubers, leaves	Asia	turnip, rutabaga, Chinese cabbage, napa cabbage, bok choy, collard greens	
	radish Raphanus sativus	roots, leaves, seed pods, seed oil, sprouting	Southeastern Asia	radish, daikon, seedpod varieties	
	carrot Daucus carota	root tubers	Persia	carrot	36.9
	parsnip Pastinaca sativa	Root tubers	Eurasia	parsnip	
	beetroot Beta vulgaris	tubers, leaves	Europe, Near East, and India	beetroot, sea beet, Swiss chard, sugar beet	
	lettuce Lactuca sativa	leaves, stems, seed oil	Egypt	lettuce, celtuce	24.9
	beans Phaseolus vulgaris Phaseolus coccineus Phaseolus lunatus	pods, seeds	Central and South America	green bean, French bean, runner bean, haricot bean, Lima bean	44.6
	broad beans Vicia faba	pods, seeds	North Africa South and southwest Asia	broad bean	

	peas Pisum sativum	pods, seeds, sprouting	Mediterranean and Middle East	pea, snap pea, snow pea, split pea	28.9
	potato Solanum tuberosum	root tubers	South America	potato	365.4
	aubergine/eggplant Solanum melongena	fruits	South and East Asia	eggplant (aubergine)	48.4
	tomato Solanum lycopersicum	fruits	South America	tomato	161.8
	cucumber Cucumis sativus	fruits	Southern Asia	cucumber	65.1
	pumpkin/squash Cucurbita spp.	fruits, flowers	Mesoamerica	pumpkin, squash, marrow, zucchini (courgette), gourd	24.6
	onion Allium cepa	bulbs, leaves	Asia	onion, spring onion, scallion, shallot	87.2
	garlic Allium sativum	bulbs	Asia	garlic	24.8
	leek Allium ampeloprasum	leaf sheaths	Europe and Middle East	leek, elephant garlic	21.7

	pepper Capsicum annuum	fruits	North and South America	pepper, bell pepper, sweet pepper	34.5
	spinach Spinacia oleracea	leaves	Central and southwestern Asia	spinach	21.7
	yam Dioscorea spp.	tubers	Tropical Africa	yam	59.5
	sweet potato Ipomoea batatas	tubers, leaves, shoots	Central and South America	sweet potato	108.0
	cassava Manihot esculenta	tubers	South America	cassava	269.1

1. Includes both carrots and turnips.
2. Productions of dry and green vegetables added up.

Nutrition and Health

Southeast Asian style stir fry ipomoea aquatica in chili and sambal

Vegetables play an important role in human nutrition. Most are low in fat and calories but are bulky and filling. They supply dietary fibre and are important sources of essential vitamins, minerals, and trace elements. Particularly important are the antioxidant vitamins A, C, and E. When vegetables are included in the diet, there is found to be a reduction in the incidence of cancer, stroke, cardiovascular disease, and other chronic ailments. Research has shown that, compared with individuals who eat less than three servings of fruits and vegetables each day, those that eat more than five servings have an approximately twenty percent lower risk of developing coronary heart disease or stroke. The nutritional content of vegetables varies considerably; some contain useful amounts of protein though generally they contain little fat, and varying proportions of vitamins such as vitamin A, vitamin K, and vitamin B_6; provitamins; dietary minerals; and carbohydrates. Vegetables contain a great variety of other phytochemicals (bioactive non-nutrient plant compounds), some of which have been claimed to have antioxidant, antibacterial, antifungal, antiviral, and anticarcinogenic properties.

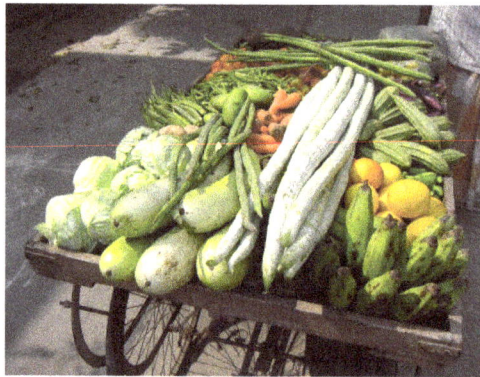

Vegetables (and some fruit) for sale on a street in Guntur, India

However, vegetables often also contain toxins and antinutrients which interfere with the absorption of nutrients. These include α-solanine, α-chaconine, enzyme inhibitors (of cholinesterase, protease, amylase, etc.), cyanide and cyanide precursors, oxalic acid, and others. These toxins are natural defenses, used to ward off the insects, predators and fungi that might attack the plant. Some beans contain phytohaemagglutinin, and cassava roots contain cyanogenic glycoside as do bamboo shoots. These toxins can be deactivated by adequate cooking. Green potatoes contain glycoalkaloids and should be avoided.

Fruit and vegetables, particularly leafy vegetables, have been implicated in nearly half the gastrointestinal infections caused by norovirus in the United States. These foods are commonly eaten raw and may become contaminated during their preparation by an infected food handler. Hygiene is important when handling foods to be eaten raw, and such products need to be properly cleaned, handled, and stored to limit contamination.

Dietary Recommendations

The USDA Dietary Guidelines for Americans recommends consuming five to nine servings of fruit and vegetables daily. The total amount consumed will vary according to age and gender, and is determined based upon the standard portion sizes typically consumed, as well as general nutritional content. Potatoes are not included in the count as they are mainly providers of starch. For most vegetables and vegetable juices, one serving is half of a cup and can be eaten raw or cooked. For leafy greens, such as lettuce and spinach, a single serving is typically a full cup. A variety of products should be chosen as no single fruit or vegetable provides all the nutrients needed for health.

International dietary guidelines are similar to the ones established by the USDA. Japan, for example, recommends the consumption of five to six servings of vegetables daily. French recommendations provide similar guidelines and set the daily goal at five servings. In India, the daily recommendation for adults is 275 grams (9.7 oz) of vegetables per day.

Production

Cultivation

Growing vegetables in South Africa

Vegetables have been part of the human diet from time immemorial. Some are staple foods but most are accessory foodstuffs, adding variety to meals with their unique flavors and at the same time, adding nutrients necessary for health. Some vegetables are perennials but most are annuals and biennials, usually harvested within a year of sowing or planting. Whatever system is used for growing crops, cultivation follows a similar pattern; preparation of the soil by loosening it, removing or burying weeds, and adding organic manures or fertilisers; sowing seeds or planting young plants; tending the crop while it grows to reduce weed competition, control pests, and provide sufficient water; harvesting the crop when it is ready; sorting, storing, and marketing the crop or eating it fresh from the ground.

Different soil types suit different crops, but in general in temperate climates, sandy soils dry out fast but warm up quickly in the spring and are suitable for early crops, while heavy clays retain moisture better and are more suitable for late season crops. The growing season can be lengthened by the use of fleece, cloches, plastic mulch, polytunnels, and greenhouses. In hotter regions, the production of vegetables is constrained by the climate, especially the pattern of rainfall, while in temperate zones, it is constrained by the temperature and day length.

Weeding cabbages in Colorado, US

On a domestic scale, the spade, fork, and hoe are the tools of choice while on commercial farms a range of mechanical equipment is available. Besides tractors, these include ploughs, harrows, drills, transplanters, cultivators, irrigation equipment, and harvesters. New techniques are changing the cultivation procedures involved in growing vegetables with computer monitoring systems, GPS locators, and self-steer programs for driverless machines giving economic benefits.

Harvesting

Harvesting beetroot in the United Kingdom

When a vegetable is harvested, it is cut off from its source of water and nourishment. It continues to transpire and loses moisture as it does so, a process most noticeable in the wilting of green leafy crops. Harvesting root vegetables when they are fully mature improves their storage life, but alternatively, these root crops can be left in the ground and harvested over an extended period. The harvesting process should seek to minimise damage and bruising to the crop. Onions and garlic can be dried for a few days in the field and root crops such as potatoes benefit from a short maturation period in warm, moist surroundings, during which time wounds heal and the skin thickens up and hardens. Before marketing or storage, grading needs to be done to remove damaged goods and select produce according to its quality, size, ripeness, and color.

Storage

All vegetables benefit from proper post harvest care. A large proportion of vegetables and perishable foods are lost after harvest during the storage period. These losses may be as high as thirty to fifty percent in developing countries where adequate cold storage facilities are not available. The main causes of loss include spoilage caused by moisture, moulds, micro-organisms, and vermin.

Temporary storage of potatoes in the Netherlands

Storage can be short-term or long-term. Most vegetables are perishable and short-term storage for a few days provides flexibility in marketing. During storage, leafy vegetables lose moisture, and the vitamin C in them degrades rapidly. A few products such as potatoes and onions have better keeping qualities and can be sold when higher prices may be available, and by extending the marketing season, a greater total volume of crop can be sold. If refrigerated storage is not available, the priority for most crops is to store high-quality produce, to maintain a high humidity level, and to keep the produce in the shade.

Proper post-harvest storage aimed at extending and ensuring shelf life is best effected by efficient cold chain application. Cold storage is particularly useful for vegetables such as cauliflower, eggplant, lettuce, radish, spinach, potatoes, and tomatoes, the optimum temperature depending on the type of produce. There are temperature-controlling technologies that do not require the use of electricity such as evaporative cooling. Storage of fruit and vegetables in controlled atmospheres with high levels of carbon dioxide or high oxygen levels can inhibit microbial growth and extend storage life.

The irradiation of vegetables and other agricultural produce by ionizing radiation can be used to preserve it from both microbial infection and insect damage, as well as from physical deterioration. It can extend the storage life of food without noticeably changing its properties.

Preservation

The objective of preserving vegetables is to extend their availability for consumption or marketing purposes. The aim is to harvest the food at its maximum state of palatability and nutritional value, and preserve these qualities for an extended period. The main causes of deterioration in vegetables after they are gathered are the actions of naturally-occurring enzymes and the spoilage caused by micro-organisms. Canning and freezing are the most commonly used techniques, and vegetables preserved by these methods are generally similar in nutritional value to comparable fresh products with regards to carotenoids, vitamin E, minerals. and dietary fiber.

Bean field and canning factory, New Jersey, US

Canning is a process during which the enzymes in vegetables are deactivated and the micro-organisms present killed by heat. The sealed can excludes air from the foodstuff to prevent subsequent deterioration. The lowest necessary heat and the minimum processing time are used in order to prevent the mechanical breakdown of the product and to preserve the flavor as far as is possible. The can is then able to be stored at ambient temperatures for a long period.

Freezing vegetables and maintaining their temperature at below −10 °C (14 °F) will prevent their spoilage for a short period, whereas a temperature of −18 °C (0 °F) is required for longer-term storage. The enzyme action will merely be inhibited, and blanching of suitably sized prepared vegetables before freezing mitigates this and prevents off-flavors developing. Not all micro-organisms will be killed at these temperatures and after thawing the vegetables should be used promptly because otherwise, any microbes present may proliferate.

Sun-drying tomatoes in Greece

Traditionally, sun drying has been used for some products such as tomatoes, mushrooms, and beans, spreading the produce on racks and turning the crop at intervals. This method suffers from several disadvantages including lack of control over drying rates, spoilage when drying is slow, contamination by dirt, wetting by rain, and attack by rodents, birds, and insects. These disadvantages can be alleviated by using solar powered driers. The dried produce must be prevented from reabsorbing moisture during storage.

High levels of both sugar and salt can preserve food by preventing micro-organisms from growing. Green beans can be salted by layering the pods with salt, but this method of preservation is unsuited to most vegetables. Marrows, beetroot, carrot, and some other vegetables can be boiled with sugar to create jams. Vinegar is widely used in food preservation; a sufficient concentration of acetic acid prevents the development of destructive micro-organisms, a fact made use of in the preparation of pickles, chutneys and relishes. Fermentation is another method of preserving vegetables for later use. Sauerkraut is made from chopped cabbage and relies on lactic acid bacteria which produce compounds that are inhibitory to the growth of other micro-organisms.

Top Producers

Farmers' market showing vegetables for sale near the Potala Palace in Lhasa, Tibet

Vegetable shop in India

Vegetables in a supermarket in the United States

Vegetables in a supermarket in Canada

In 2010, China was the largest vegetable producing nation, with over half the world's production. India, the United States, Turkey, Iran, and Egypt were the next largest producers. China had the highest area of land devoted to vegetable production, while the highest average yields were obtained in Spain and the Republic of Korea.

Country	Area cultivated thousand hectares (2,500 acres)	Yield thousand kg/ha (89 lbs/acre)	Production thousand tonnes (1,100 short tons)
China	23,458	230	539,993
India	7,256	138	100,045
United States	1,120	318	35,609
Turkey	1,090	238	25,901
Iran	767	261	19,995
Egypt	755	251	19,487
Italy	537	265	14,201
Russia	759	175	13,283
Spain	348	364	12,679
Mexico	681	184	12,515
Nigeria	1844	64	11,830
Brazil	500	225	11,233
Japan	407	264	10,746
Indonesia	1082	90	9,780
South Korea	268	364	9,757
Vietnam	818	110	8,976
Ukraine	551	162	8,911
Uzbekistan	220	342	7,529
Philippines	718	88	6,299
France	245	227	5,572
Total world	55,598	188	1,044,380

Standards

The International Organization for Standardization (ISO) sets international standards to ensure that products and services are safe, reliable, and of good quality. There are a number of ISO standards regarding fruits and vegetables. ISO 1991-1:1982 lists the botanical names of sixty-one species of plants used as vegetables along with the common names of the vegetables in English, French, and Russian. ISO 67.080.20 covers the storage and transport of vegetables and their derived products.

Murcia

Murcia is a city in south-eastern Spain, the capital and most populous city of the Autonomous Community of the Region of Murcia, and the seventh largest city in the country, with a population of 442,573 inhabitants in 2009 (about one third of the total population of the Region). The population of the metropolitan area was 689,591 in 2010. It is located on the Segura River, in the Southeast of the Iberian Peninsula, noted by a climate with hot summers, mild winters, and relatively low precipitation.

Murcia was founded by the emir of Cordoba Abd ar-Rahman II in 825 with the name Mursiyah مرسية and nowadays is mainly a services city and a university town. Highlights for visitors include the Cathedral of Murcia and a number of baroque buildings, renowned local cuisine, Holy Week procession works of art by the famous Murcian sculptor Francisco Salzillo, and the Fiestas de Primavera (Spring Festival).

The city, as the capital of the comarca Huerta de Murcia is called Europe's orchard due to its long agricultural tradition and its fruit, vegetable, and flower production and exports.

Geography

Murcia is located near the center of a low-lying fertile plain known as the huerta (orchard or vineyard) of Murcia. The Segura River and its right-hand tributary, the Guadalentín, run through the area. The city has an elevation of 43 metres (141 ft) above sea level and covers approximately 882 square kilometres (341 sq mi).

The best known and most dominant aspect of the municipal area's landscape is the orchard. In addition to the orchard and urban zones (Alfonso X, Gran Via, Jaime I, and others), the great expanse of the municipal area is made up of different landscapes: badlands, groves of Carrasco pine trees in the precoastal mountain ranges and, towards the south, a semi-steppe region.

A large regional park, the Parque Regional de Carrascoy y el Valle, lies just to the south of the city.

Segura River

Murcia is located in the Segura valley

The Segura River crosses an alluvial plain (Vega Media del Segura), part of a Mediterranean pluvial system. The river crosses the city from west to east. Its volumetric flow is mostly small but the river is known to produce occasional flooding, like those that inundated the capital in 1946, 1948, 1973 or 1989. The Segura was recognized as one of the most polluted rivers in Europe.

Mountains and Hills

The Segura river's Valley is surrounded by two mountain ranges, the hills of Guadalupe, Espinardo, Cabezo de Torres, Esparragal and Monteagudo in the north and the Cordillera Sur in the south. The municipality itself is divided into southern and northern zones by a series of mountain ranges, the aforementioned Cordillera Sur (Carrascoy, El Puerto, Villares, Columbares, Altaona, and Escalona). These two zones are known as Field of 'Murcia (in the south of Cordillera Sur) and Orchard of Murcia (the Segura Valley in the north of Cordillera Sur). Near the plain's center, the steep hill of Monteagudo protrudes dramatically.

Districts

The 881.86-square-kilometre (340.49 sq mi) territory of Murcia's municipality is made up of 54 pedanías (suburban districts) and 28 barrios (city neighbourhood districts). The barrios make up the 12.86-square-kilometre (4.97 sq mi) the main urban portion of the city. The historic city center is approximately 3 square kilometres (1 sq mi) of the urbanized downtown portion of Murcia.

Climate

Murcia has a hot semi-arid climate (Köppen climate classification: BSh), with arid climate (BWh) influences. Given its proximity to the Mediterranean Sea, it has mild winters and hot summers.

It averages more than 320 days of sun per year. Occasionally, Murcia has heavy rains where the precipitation for the entire year will fall over the course of a few days.

In the coldest month, January, the average temperature range is a high of 16.6 °C (62 °F) during the day and a low of 4.7 °C (40 °F) at night. In the warmest month, August, the range goes from 34.2 °C (94 °F) during the day to 20.9 °C (70 °F) at night. Temperatures almost always reach or exceed 40 °C (104 °F) on at least one or two days per year. In fact, Murcia holds temperature records close to the highest recorded in southern Europe since reliable meteorological records commenced in 1950. The official record for Murcia stands at a stifling 47.2 °C (117.0 °F), at Alcantarilla airport in the western suburbs on July 4, 1994 with 45.7 °C (114.3 °F) being recorded at a station near the city centre on the same day.

Climate data for Murcia (1981–2010)													
Month	Jan	Feb	Mar	Apr	May	Jun	Jul	Aug	Sep	Oct	Nov	Dec	Year
Record high °C (°F)	25.8 (78.4)	29.4 (84.9)	32.6 (90.7)	37.4 (99.3)	38.5 (101.3)	42.5 (108.5)	47.2 (117)	43.2 (109.8)	44.6 (112.3)	34.9 (94.8)	31.0 (87.8)	25.8 (78.4)	47.2 (117)
Average high °C (°F)	16.6 (61.9)	18.4 (65.1)	20.9 (69.6)	23.3 (73.9)	26.6 (79.9)	31.0 (87.8)	34.0 (93.2)	34.2 (93.6)	30.4 (86.7)	25.6 (78.1)	20.3 (68.5)	17.2 (63)	24.9 (76.8)
Daily mean °C (°F)	10.6 (51.1)	12.2 (54)	14.3 (57.7)	16.5 (61.7)	20.0 (68)	24.2 (75.6)	27.2 (81)	27.6 (81.7)	24.2 (75.6)	19.8 (67.6)	14.6 (58.3)	11.5 (52.7)	18.6 (65.5)
Average low °C (°F)	4.7 (40.5)	5.9 (42.6)	7.7 (45.9)	9.7 (49.5)	13.3 (55.9)	17.4 (63.3)	20.3 (68.5)	20.9 (69.6)	18.0 (64.4)	13.9 (57)	8.9 (48)	5.8 (42.4)	12.3 (54.1)
Record low °C (°F)	−7.5 (18.5)	−3.9 (25)	−2.4 (27.7)	0.0 (32)	4.0 (39.2)	8.0 (46.4)	13.0 (55.4)	14.0 (57.2)	9.6 (49.3)	4.4 (39.9)	−1.0 (30.2)	−6.0 (21.2)	−7.5 (18.5)
Average precipitation mm (inches)	27 (1.06)	27 (1.06)	30 (1.18)	25 (0.98)	28 (1.1)	18 (0.71)	3 (0.12)	8 (0.31)	32 (1.26)	36 (1.42)	32 (1.26)	29 (1.14)	297 (11.69)
Average precipitation days (≥ 1 mm)	4	4	3	4	4	2	1	1	3	4	4	4	37
Mean monthly sunshine hours	189	190	223	256	289	323	353	317	239	217	186	172	2,967
Source: Agencia Estatal de Meteorología													

Climate data for Murcia—San Javier (Airport 4 m, near sea) (1981–2010)												
Month	Jan	Feb	Mar	Apr	May	Jun	Jul	Aug	Sep	Oct	Nov	Dec	Year
Record high °C (°F)	26.2 (79.2)	27.8 (82)	30.0 (86)	32.0 (89.6)	34.5 (94.1)	36.9 (98.4)	40.5 (104.9)	40.0 (104)	39.4 (102.9)	35.5 (95.9)	30.0 (86)	27.0 (80.6)	40.5 (104.9)
Average high °C (°F)	16.0 (60.8)	16.7 (62.1)	18.5 (65.3)	20.4 (68.7)	22.9 (73.2)	26.4 (79.5)	28.9 (84)	29.5 (85.1)	27.5 (81.5)	24.0 (75.2)	19.8 (67.6)	17.6 (63.7)	22.3 (72.1)
Daily mean °C (°F)	10.8 (51.4)	11.6 (52.9)	13.4 (56.1)	15.3 (59.5)	18.4 (65.1)	22.2 (72)	24.8 (76.6)	25.5 (77.9)	23.2 (73.8)	19.4 (66.9)	14.9 (58.8)	11.9 (53.4)	17.6 (63.7)
Average low °C (°F)	5.5 (41.9)	6.5 (43.7)	8.4 (47.1)	10.2 (50.4)	13.8 (56.8)	17.9 (64.2)	20.7 (69.3)	21.5 (70.7)	18.9 (66)	14.7 (58.5)	10.0 (50)	6.8 (44.2)	12.9 (55.2)
Record low °C (°F)	−3.8 (25.2)	−4.0 (24.8)	−3.0 (26.6)	1.0 (33.8)	4.8 (40.6)	9.5 (49.1)	11.0 (51.8)	12.0 (53.6)	7.9 (46.2)	4.0 (39.2)	−1.5 (29.3)	−5.4 (22.3)	−5.4 (22.3)
Average precipitation mm (inches)	42 (1.65)	27 (1.06)	24 (0.94)	23 (0.91)	25 (0.98)	7 (0.28)	2 (0.08)	7 (0.28)	39 (1.54)	39 (1.54)	47 (1.85)	30 (1.18)	313 (12.32)
Average precipitation days (≥ 1 mm)	4	3	3	3	3	1	0	1	3	4	4	4	33
Mean monthly sunshine hours	173	171	206	224	266	288	307	283	224	200	162	156	2,621
Source: Agencia Estatal de Meteorología												

History

Muslim architecture of the Alcázar Seguir in Santa Clara Museum inside of Monasterio de Santa Clara la Real.

It is widely believed that Murcia's name is derived from the Latin words of Myrtea or Murtea, meaning land of Myrtle (the plant is known to grow in the general area), although it may also be a derivation of the word Murtia, which would mean Murtius Village (Murtius was a common Roman name). Other research suggests that it may owe its name to the Latin Murtae (Mulberry), which covered the regional landscape for many centuries. The Latin name eventually changed into the Arabic Mursiya, and then, Murcia.

Entrance of James I of Aragon at Murcia in 1266.

The city in its present location was founded with the name Madinat Mursiyah (city of Murcia) in AD 825 by Abd ar-Rahman II, who was then the emir of Córdoba. Muslim planners, taking advantage of the course of the river Segura, created a complex network of irrigation channels that made the town's agricultural existence prosperous. In the 12th century the traveler and writer Muhammad al-Idrisi described the city of Murcia as populous and strongly fortified. After the fall of the Caliphate of Cordoba in 1031, Murcia passed under the successive rules of the powers seated variously at Almería, Toledo and Seville. After the fall of Almoravide empire, Muhammad Ibn Mardanis made Murcia the capital of an independent kingdom. At this time, Murcia was a very prosperous city, famous for its ceramics, exported to Italian towns, as well as for silk and paper industries, the first in Europe. The coinage of Murcia was considered as model in all the continent. The mystic Ibn Arabi (1165–1240) and the poet Ibn al-Jinan (d.1214) were born in Murcia during this period.

In 1172 Murcia was conquered by the north African based Almohades, the last Muslim empire to rule southern Spain, and as the forces of the Christian Reconquista gained the upper hand, was the capital of a small Muslim emirate from 1223 to 1243. By the treaty of Alcaraz, in 1243, the Christian king Ferdinand III of Castile made Murcia a protectorate, getting access to the Mediterranean sea while Murcia was protected against Granada and Aragon. The Christian population of the town became the majority as immigrants poured in from almost all parts of the Iberian Peninsula. Christian immigration was encouraged with the goal of establishing a loyal Christian base. These measures led to the Muslim population revolt in 1264, which was quelled by James I of Aragon in 1266, bringing Aragonese and Catalonian immigrants with him.

After this, during the reign of Alfonso X of Castile, Murcia was one of his capitals with Toledo and Seville.

The Murcian duality: Catalonian population in a Castillian territory, brought the subsequent conquest of the city by James II of Aragon in 1296. In 1304, Murcia was finally incorporated into Castile under the Treaty of Torrellas.

Murcia Flood in 1879

Murcia's prosperity declined as the Mediterranean lost trade to the ocean routes and from the wars between the Christians and the Ottoman Empire. The old prosperity of Murcia became crises during 14th century because of its border location with the neighbouring Muslim kingdom of Granada, but flourished after its conquest in 1492 and again in the 18th century, benefiting greatly from a boom in the silk industry. Most of the modern city's landmark churches, monuments and old architecture date from this period. In this century, Murcia lived an important role in Bourbon victory in the War of the Spanish Succession, thanks to the Cardinal Belluga. In 1810, Murcia was looted by Napoleonic troops; it then suffered a major earthquake in 1829. According to contemporaneous accounts, an estimated 6,000 people died from the disaster's effects across the province. Plague and cholera followed.

The town and surrounding area suffered badly from floods in 1651, 1879, and 1907, though the construction of a levee helped to stave off the repeated floods from the Segura. A popular pedestrian walkway, the Malecon, runs along the top of the levee.

Murcia has been the capital of the province of Murcia since 1833 and, with its creation by the central government in 1982, capital of the autonomous community (which includes only the city and the province). Since then, it has become the seventh most populated municipality in Spain, and a thriving services city.

On May 11, 2011, the city of Lorca and surrounding area was struck by a 5.3 magnitude earthquake. At least 4 people were reported to have died as a result of the earthquake.

Demographics

Murcia Cathedral of Santa Maria

Al-Andalusian palatial complex and neighborhood of San Esteban

The town hall

Murcia has 433,850 inhabitants (INE 2008) making it the seventh-largest Spanish municipality by population. When adding in the municipalities of Alcantarilla, Alguazas, Beniel, Molina de Segura, Santomera, and Las Torres de Cotillas, the metropolitan area has 564,036 inhabitants making it the twelfth most populous metropolitan area in

Spain. Nevertheless, due to Murcia's large municipal territory, its population density (472 hab./km², 760 hab./sq.mi.) does not likewise rank among Spain's highest.

According to the official population data of the INE, 10% of the population of the municipality reported belonging to a foreign nationality as of 2005.

The majority of the population identify as Christian. There is also a sizeable Muslim population as well as a growing Jewish community.

Main Sights

The Cathedral of Murcia was built between 1394 and 1465 in the Castilian Gothic style. Its tower was completed in 1792 and shows a blend of architectural styles. The first two stories were built in the Renaissance style (1521–1546), while the third is Baroque. The bell pavilion exhibits both Rococo and Neoclassical influences. The main façade (1736–1754) is considered a masterpiece of the Spanish Baroque style.

Other noteworthy buildings in the square shared by the Cathedral (Plaza Cardinal Belluga) are the colorful Bishop's Palace (18th century) and a controversial extension to the town hall by Rafael Moneo (built in 1999).

The Glorieta, which lies on the banks of the Segura River, has traditionally been the center of the town. It is a pleasant, landscaped city square that was constructed during the 18th century. The ayuntamiento (city hall) of Murcia is located in this square.

Pedestrian areas cover most of the old town of the city, which is centered around Platería and Trapería Streets. Trapería goes from the Cathedral to the Plaza de Santo Domingo, formerly a bustling market square. Located in Trapería is the Casino, a social club erected in 1847, with a sumptuous interior that includes a Moorish-style patio inspired by the royal chambers of the Alhambra near Granada. The name Plateria refers to plata (silver), as this street was the historical focus for the commerce of rare metals by Murcia's Jewish community. The other street, Traperia, refers to trapos, or cloths, as this was once the focus for the Jewish community's garment trade.

Several bridges of different styles span the river Segura, from the Puente de los Peligros, eighteenth century stone bridge with a Lady chapel on one of its sides; to modern bridges designed by Santiago Calatrava or Javier Manterola; through others such as the Puente Nuevo, an iron bridge of the early twentieth century

Other notable places around Murcia include:

- Santa Clara monastery, a Gothic and Baroque monument where is located a museum with the Moorish palace's remains from the 13th century, called Alcázar Seguir.

- The Malecón boulevard, a former retaining wall for the Río Segura's floods.

- La Fuensanta sanctuary and adjacent El Valle regional park

- Los Jerónimos monastery (18th century)

- Romea theatre (19th century)

- Almudí Palace (17th century), a historic building with coats of arms on its façade. On its interior there are Tuscan columns, and since 1985 it hosts the city archives and usually houses exhibitions.

- Monteagudo Castle (11th century)

- Salzillo Museum

- San Juan de Dios church-museum, Baroque and Rococo circular church with the remains of the Moorish palace mosque from the 12th century in the basament, called Alcázar Nasir.

In the metropolitan area are also the Azud de la Contraparada reservoir and the Noria de La Ñora water wheel.

Festivals

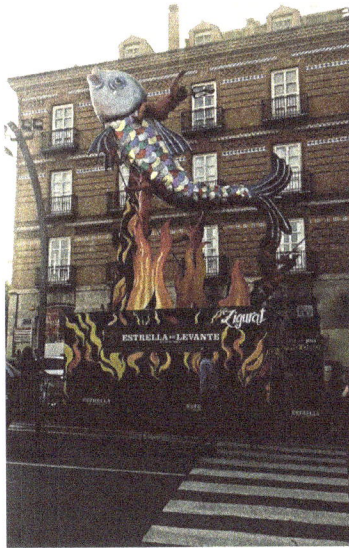

The Burial of the Sardine in Murcia

The Holy Week procession hosted by the city is among the most famous throughout Spain. This traditional festival portrays the events which lead up to and include the Crucifixion according to the New Testament. Life-sized, finely detailed sculptures by Francisco Salzillo (1707–1783) are removed from their museums and carried around the city in elegant processions amid flowers and, at night, candles, pausing at stations which are meant to re-enact the final moments before the crucifixion of Jesus.

The most colorful festival in Murcia may come one week after Holy Week, when locals dress up in traditional huertano clothing to celebrate the Bando de la Huerta (Orchard parade) on Tuesday and fill the streets for The Burial of the Sardine in Murcia. parade the following Saturday. This whole week receives the name of Fiestas de Primavera (Spring Fest).

Murcia's Three Cultures International Festival happens each May and was first organized with the intent of overcoming racism and xenophobia in the culture. The festival seeks to foster understanding and reconciliation between the three cultures that have cohabited the peninsula for centuries, if not millennia: Christians, Jews and Muslims. Each year, the festival celebrates these three cultures through music, exhibitions, symposiums and conferences.

Economy

Trapería Street in the old town

Economically, Murcia predominantly acts as a centre for agriculture and tourism. It is common to find Murcia's tomatoes and lettuce, and especially lemons and oranges, in European supermarkets. Murcia is a producer of wines, with about 40,000 hectares (100,000 acres) devoted to grape vineyards. Most of the vineyards are located in Ricote and Jumilla. Jumilla is a plateau where the vineyards are surrounded by mountains.

Murcia has some industry, with foreign companies choosing it as a location for factories, such as Henry Milward & Sons (which manufactures surgical and knitting needles) and American firms like General Electric and Paramount Park Studios.

During the 2000s, the economy of the region turned towards "residential tourism" in which people from northern European countries have a second home in the area. Europeans and Americans are able to learn Spanish in the academies in the town center.

The economy of Murcia is supported by fairs and congresses, museums, theatres, cinema, music, aquariums, bullfighting, restaurants, hotels, camping, sports, foreign students, and tourism.

Transportation

Tram of Murcia.

By Plane

Murcia-San Javier Airport (MJV) is located on the edge of the Mar Menor close to the town of San Javier, 45 kilometres (28 miles) southeast of Murcia. There is also an airport at the neighboring city of Alicante 70 km (43 miles) from Murcia. Furthermore, there is a new airport in development to be located in the town of Corvera, 23 km (14 miles).

By Bus

Bus service is provided by LatBus, which operates the interurban services. Urban bus services is offered by a new operator, TM(Transportes de Murcia), an UTE (Joint Venture) formed by Ruiz, Marín & Fernanbús.

By Tram

Tramways are managed by Tranvimur. As of 2007, 2 kilometres (1 mile) of line were available. Since 2011, one line is connecting the city center (Plaza Circular) with the University Campus and the Football Stadium.

By Train

Train connections are provided by RENFE. Murcia has a railway station called Murcia del Carmen, located in the neighborhood of the same name. Several long-distance lines link the city with Madrid, through Albacete, as well as Valencia, and Cataluña up to Montpellier in France. Murcia is also the center of a local network. The line C-1 connects the city to Alicante, and the line C-2 connects Murcia to Alcantarilla, Lorca and Águilas. It also has two regional lines connecting it to Cartagena and Valencia.

Healthcare

The hospitals and other public primary healthcare centers belong to the Murcian Healthcare Service. There are three public hospitals in Murcia:

- Ciudad Sanitaria Virgen de La Arrixaca in El Palmar that includes obstetrics and paediatrics units

- Hospital Reina Sofía

- Hospital Morales Meseguer

Education

University of Murcia

University of Murcia (cloister)

Murcia has two universities:

- one public university: the University of Murcia, founded in 1272.

- one private university: the Catholic University Saint Anthony, founded in 1996.

There are several high schools, elementary schools, and professional schools. Murcia has three types of schools for children: private schools such as El Limonar International School, Murcia (an American international school) and King's College, Murcia (a British international school), semi-private schools (concertado), which are private schools that receive government funding and sometimes offer religious instruction, and public schools such as Colegio Publico (CP) San Pablo or the centenary CP Cierva Peñafiel, one of the oldest ones. The French international school, Lycée Français André Malraux de Murcie, is in nearby Molina de Segura.

The private schools and concertados can be religious (Catholic mostly but any religion is acceptable) or secular, but the public schools are strictly secular. Concertado or semi-private or quasi-private schools fill a need by providing schools where the government isn't able to or predate the national school system.

Instituto Licenciado Cascales is one of the oldest in the city, built in 1724, and perhaps the most emblematic. IES Alfonso X El Sabio is the only school in Murcia to offer the prestigious International High School Diploma.

Murcia also offers Adult Education for people who want to return to complete high school and possibly continue on to the university.

Sports Teams

- Real Murcia: Spanish Third Division football
- CF Atlético Ciudad: Spanish Third Division (Group 2) football—dissolved in 2010
- CB Murcia: Liga ACB basketball
- ElPozo Murcia Turística FS: futsal
- The Hispania Racing F1 Team is also based in Murcia, and receives sponsorship from the tourist board
- CAV Murcia 2005: Superliga Femenina de Voleibol volleyball
- Origen (esports): League of Legends team

Twin Towns—Sister Cities

Murcia is twinned with:

- Lecce, Italy
- Grasse, France
- Irapuato, Mexico
- Murcia, Philippines
- Łódź, Poland, since 1999

Tenbury Wells

Tenbury Wells (locally Tenbury) is a market town and civil parish in the north-western extremity of the Malvern Hills District of Worcestershire, England, which at the 2011 census had a population of 3,777.

Geography

Tenbury Wells lies on the south bank of the River Teme, which forms the border between Shropshire and Worcestershire. It is in the north-west of the Malvern Hills District. The settlement of Burford in Shropshire lies on the north bank of the river.

History

From 1894 to 1974, it was a rural district, comprising itself and villages such as Stoke Bliss, Eastham and Rochford. From 1974 Tenbury was in the District of Leominster until it became part Malvern Hills District when Leominster District Council was taken over by Herefordshire Council in April 1998.

The history of Tenbury Wells extends as far back as the Iron Age. The town is often thought of as the home to the Castle Tump, but this is now in Burford, Shropshire due to boundary changes. Though the Tump, possibly the remains of an early Norman motte and bailey castle, can be seen from the main road (A456) there are no visible remains of the castle that was constructed to defend and control the original River Teme crossing. It has also been described as "... the remains of an 11th century Norman Castle."

A legal record of 1399 mentions a place spelt perhaps as Temedebury which may be a further variation in spelling.

Tenbury was in the upper division of Doddingtree Hundred.

Originally named "Temettebury", the town was granted a Royal Charter to hold a market in 1249. Over time, the name changed to "Tenbury", and then added the "Wells" following the discovery of mineral springs and wells in the town in the 1840s. The name of the railway station, which was on the now-defunct Tenbury & Bewdley Railway, was changed in 1912, in an attempt to publicise the mineral water being produced from the wells around the town.

The St Michael and All Angels Choir School devoted to the Anglican choral tradition by Frederick Ouseley closed in 1985 and the buildings now serve alternative educational purposes.

For over 100 years Tenbury has been well known throughout the country for its winter auctions of holly and mistletoe (and other Christmas products). It is also known for its "Chinese-gothic" Pump Room buildings, built in 1862, which re-opened in 2001, following a major restoration. They are now owned by Tenbury Town Council, having been transferred from Malvern Hills District Council in September 2008.

Architecture

Eastham Bridge near Tenbury, which collapsed in May 2016

One notable architectural feature in the town is the unique (often described as Chinese-Gothic) Pump Rooms, designed by James Cranston in the 1860s, to house baths where the mineral water was available.

Other notable structures in Tenbury include the parish church of St Mary with a Norman tower, and a number of monuments. The church was essential rebuilt by Henry Woodyer between 1864 and 1865.

The part-medieval bridge over the River Teme, linking Tenbury to Burford, Shropshire was rebuilt by Thomas Telford following flood damage in 1795.

The Grade II-listed Eastham bridge dramatically collapsed into the River Teme on 24 May 2016. There were no reports of any casualties.

The Victorian Workhouse, designed by George Wilkinson, was used as the local Council Buildings from 1937 to the early 21st century and is currently being converted into residential housing. The Victorian infirmary behind the workhouse is scheduled to be demolished to create car parking for a new Tesco Superstore.

The unique Victorian corrugated iron isolation hospital was demolished on 24 October 2006.

Local Interest

Markets

Markets are held on Tuesday mornings, Friday mornings, and Saturday mornings, in and around the town's Round Market building, which was built by James Cranston in 1858. In 2013, a new monthly 'local producers market' started, initially held near the Pump Rooms, more recently on Teme Street.

Apple and Fruit Heritage

Tenbury was also known as "the town in the orchard" due to the large numbers of fruit orchards of apple trees and also pears, quince and plum trees, in the immediate vicinity of the town. This heritage is revisited every October during the Tenbury Applefest. Tenbury Applefest website.

Tenbury in Poetry

Orchards gay with blossom,
Beauty, there to see,
Hollows where breeze is tender,
Moorlands where wind breaks free;
Sowing, Lambing, and Harvest,
Overlooked by Giant Clee,
Hop Kilns, Farmsteads, and TENBURY,
This is happiness for me;

Power Station Shelved

A proposal to build a biomass power station on a business park failed due to residents' concern about the disruption to local businesses during its construction. The proposal continued to attract protests, and in July 2007 a petition against the plans was signed by more than 2,300 people. In July 2009 it was announced that the £965,000 grant offered to the power station had been withdrawn and the project shelved.

Local Flooding

For several centuries Tenbury has been subject to regular flooding on many occasions, and most recently in 2007 and in 2008. The first flood was caused by the River Teme and the Kyre Brook bursting their banks. The second was caused by a combination of 15mm (0.59 in) of rain falling in an hour and the town's drainage system (much of which was blocked) failing to cope, creating flash flooding. The third flood again involved the River Teme and the Kyre Brook bursting their banks. The 2008 flood damage was caused by a combination of the drainage not having been upgraded since the 2007 floods and the wall on Market Street (which should hold back the Kyre Brook) not having been rebuilt following the 2007 floods. Since then much work has been done in respect of improved drainage and particularly defences in Market Street.

Regal Cinema

The Regal Cinema on Teme Street in Tenbury Wells opened in 1937. It operated as a commercial cinema as one of six in the Craven Cinemas chain, until the decline of British cinema led to its closure in 1966. Following purchase by Tenbury Town Council to prevent demolition, various volunteer groups have run it.

The Regal has been subject of a Heritage Lottery Fund supported restoration project. Replicas of the 1930s mediterranean murals by artist George Legge have been painted around the auditorium, the detailing on the front of the building has been recreated, and neon lighting has been erected on the front canopy. The building, owned by Tenbury Town Council is now under the management of a trust. Modern equipment now allows the showing of recently released films, live broadcasts and live acts. Paul Daniels was its patron until his death.

In 2016 The Regal in Tenbury has been nominated for the "Britain Has Spirit" award. An award that could see The Regal win £1,000 to host a street party for the Tenbury community if they win the regional vote, and possibly £25,000 if they win the national vote. The Regal are currently in the public voting stage and members of the public have up until the 16th June 2016 to vote for them to win. The competition is being run by Together Mutual Insurance

Education

For primary education Tenbury Wells is served by Tenbury CofE Primary School on Bromyard Road. Tenbury High Ormiston Academy on Oldwood Road is the main secondary school for the area, while King's St Michael's College (also on Oldwood Road) is an independent international boarding school.

Nearest Railway Stations

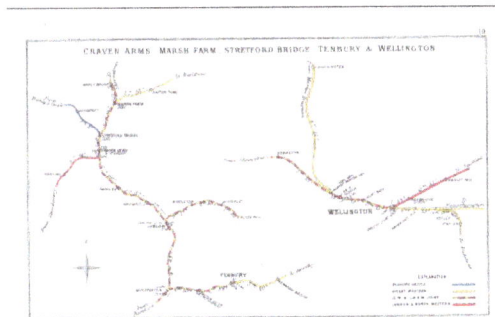

Railway Clearing House Junction Diagram of 1903. Woofferton railway station on the Welsh Marches Line.

The nearest open stations are located on the Welsh Marches Line are Ludlow railway station and Leominster.

The nearest point of operational railway is at Woofferton railway station, but it is currently closed.

Orchard House

Orchard House is a historic house museum in Concord, Massachusetts, USA. It was the longtime home of Amos Bronson Alcott (1799–1888) and his family, including his daughter Louisa May Alcott (1832-1888) who wrote and set her novel Little Women (1868–69) there.

History

The Alcotts had first moved to Concord in 1840, although they left in 1843 to start Fruitlands, a utopian agrarian commune in nearby Harvard. The family returned in 1845 and purchased a house named "Hillside", but left again in 1852, selling to Nathaniel Hawthorne who renamed it The Wayside.

The Alcotts returned to Concord once again in 1857. They moved into Orchard House, which was then two-story clapboard farmhouse, in the spring of 1858. At the time of purchase the site included two early eighteenth-century houses on a 12-acre (49,000m²) apple orchard. Consequently, the Alcotts named it Orchard House. "'Tis a pretty retreat", Bronson Alcott wrote soon after moving in, "and ours; a family mansion to take pride in, rescued as it is from deformity and disgrace".

A. Bronson Alcott made significant changes to the building. He installed alcoves for busts retrieved from his failed Temple School, repaired the staircase, installed bookcases, constructed a back studio for his youngest daughter May's artwork, and installed a rustic fence around the property. He also moved a smaller tenant house to adjoin the rear of the main house, making a single larger structure. While the home was being renovated, the family rented rooms next door at The Wayside while the Hawthornes were living in England. Later, Lydia Maria Child visited the house and recorded her thoughts: "The result is a house full of queer nooks and corners and all manner of juttings in and out. It seems as if the spirit of some old architect had brought it from the Middle Ages and dropped it down in Concord... The whole house leaves a general impression of harmony, of a medieval sort".

Orchard House is adjacent to The Wayside on the historic "American Mile" roadway toward Lexington, and is less than a half-mile from Bush the home of Ralph Waldo Emerson, where Henry David Thoreau and the Alcotts were frequent visitors.

The Alcotts in Residence

Orchard House was the most permanent home of the Alcotts, with the family in residence from 1858 to 1877. During this period, the family included Bronson, his wife Abigail May, and their daughters Anna, Louisa, and May. Elizabeth, the model for Beth March, had died in March 1858, just weeks before the family moved in.

Orchard House, 1941

The Alcott girls befriended the Hawthorne children, who lived next door, though Nathaniel Hawthorne himself was elusive. Bronson Alcott was disappointed and recorded, "Nobody gets a chance to speak with him unless by accident." However, he added, "Still he has a tender kindly side, and a voice that a woman might own, the hesitance is so taking, and the tones so remote from what you expected."

The Alcotts were vegetarians and harvested fruits and vegetables from the gardens and orchard on the property. Conversations about abolitionism, women's suffrage and social reform were often held around the dining room table. The family performed theatricals using the dining room as their stage while guests watched from the adjoining parlor.

The parlor was a formal room with arched niches built by Bronson to display busts of his favorite philosophers, Socrates and Plato. On May 23, 1860, Anna married John Bridge Pratt in this room.

The youngest daughter, May, was a talented artist. Her bedroom contains sketches of angelic, mythological and biblical figures on the woodwork and doors. In Louisa's room, May painted a panel of calla lilies as well as an owl on the fireplace. Copies of Turner landscapes by May adorn various rooms in Orchard House.

In 1868, Louisa May wrote her beloved classic novel Little Women in her room on a special "shelf desk" built by her father. Set within the house, its characters are based on members of her family, with the plot loosely based on the family's earlier years and events that transpired at The Wayside. Also written in the house were Bronson Alcott's

Ralph Waldo Emerson (1865; published 1882), Tablets (1868), Concord Days (1872), and Table Talk (1877).

On the grounds to the west of the house is a structure designed and built by Bronson Alcott originally known as "Hillside Chapel" and later as "The Concord School of Philosophy". Operating from 1879 to 1888, the School was one of the first highly successful adult education centers in the country.

In 1877, Louisa May Alcott bought the a home on Main Street for her sister Anna. After Mrs. Alcott's death in the same year, Louisa and her father moved into the home as well. Orchard House was then sold to long-time family friend William Torrey Harris in 1884.

The Orchard House Today

Orchard House is open for guided tours daily, with the exceptions of Easter, Thanksgiving, Christmas, and January 1 and 2. An admission fee is charged.

The exterior looks much as it did in the Alcotts' day. Care has been taken to keep extensive structural preservation work invisible. All of the furnishings are original to the mid-nineteenth century, about 75% belonged to the Alcott family, and the rooms look very much as they did when the Alcotts were in residence.

The Hillside Chapel

The dining room contains family china, portraits of the family members, and paintings by May along with period furnishings. The parlor is decorated with period wallpaper and a patterned reproduction carpet while family portraits and watercolors by May adorn the walls. Abigail May's bread board, mortar and pestle, tin spice chest and wooden bowls are displayed on the hutch table in the kitchen. Other original kitchen features include a laundry drying rack designed by Bronson, and a soapstone sink bought by Louisa. The study is furnished with Bronson's library table, chair and desk. The parent's bedroom contains many of Abigail May's possessions, including photographs, furniture, and hand made quilts.

Orchard House has continued the tradition of Mr. Alcott's Concord School of Philosophy by hosting "The Summer Conversational Series" since 1977, and has recently added a "Teacher Institute" component. The Hillside Chapel is also used for youth programs, poetry readings, historical reenactments, and other special events.

Fruita, Utah

Fruita is the best-known settlement in Capitol Reef National Park in Wayne County, Utah, United States. It is located at the confluence of Fremont River and Sulphur Creek.

History

Fruita was established in 1880 by a group of Mormons led by Nels Johnson, under the name "Junction." The town became known as Fruita in 1902 or 1904. In 1900, Fruita was named The Eden of Wayne County for its large orchards. Fruita was abandoned in 1955 when the National Park Service purchased the town to be included in Capitol Reef National Park.

Today few buildings remain, except for the restored schoolhouse and the Gifford house and barn. The orchards remain, now under the ownership of the National Park Service, and have about 2,500 trees. The orchards are preserved by the NPS as a "historic landscape" and a small crew takes care of them by pruning, irrigating, replanting, and spraying them.

The one-room schoolhouse was built and opened in 1896. The few students were instructed mainly in reading, writing, and arithmetic, but when the teachers were capable, they also studied other subjects such as history or geography. The room was also used for balls and religious services. It was renovated in 1966 by the National Park Service.

Fruita is currently the heart and administrative center of Capitol Reef National Park.

Fabaceae

This article is about Fabaceae s.l. (or Leguminosae), as defined by the APG System. For Fabaceae s.s. (or Papilionaceae), as defined by less modern systems.

The Fabaceae, Leguminosae or Papilionaceae, commonly known as the legume, pea, or bean family, are a large and economically important family of flowering plants. It includes trees, shrubs, and perennial or annual herbaceous plants, which are easily recognized by their fruit (legume) and their compound, stipulated leaves. The family is widely distributed, and is the third-largest land plant family in terms of number of species, behind only the Orchidaceae and Asteraceae, with about 751 genera and some 19,000 known species . The five largest of the genera are Astragalus (over 3,000 species), Acacia (over 1000 species), Indigofera (around 700 species), Crotalaria (around 700 species) and Mimosa (around 500 species), which constitute about a quarter of all legume species. The ca. 19,000 known legume species amount to about 7% of flowering plant species. Fabaceae is the most common family found in tropical rainforests and in dry forests in the Americas and Africa.

Recent molecular and morphological evidence supports the fact that the Fabaceae is a single monophyletic family. This point of view has been supported not only by the degree of interrelation shown by different groups within the family compared with that found among the Leguminosae and their closest relations, but also by all the recent phylogenetic studies based on DNA sequences. These studies confirm that the Fabaceae are a monophyletic group that is closely related to the Polygalaceae, Surianaceae and Quillajaceae families and that they belong to the order Fabales.

Along with the cereals, some fruits and tropical roots a number of Leguminosae have been a staple human food for millennia and their use is closely related to human evolution.

A number are important agricultural and food plants, including Glycine max (soybean), Phaseolus (beans), Pisum sativum (pea), Cicer arietinum (chickpeas), Medicago sativa (alfalfa), Arachis hypogaea (peanut), Lathyrus odoratus (sweet pea), Ceratonia siliqua (carob), and Glycyrrhiza glabra (liquorice). A number of species are also weedy pests in different parts of the world, including: Cytisus scoparius (broom), Robinia pseudoacacia (black locust), Ulex europaeus (gorse), Pueraria lobata (kudzu), and a number of Lupinus species.

Description

The fruit of Gymnocladus dioicus

Fabaceae range in habit from giant trees (like Koompassia excelsa) to small annual herbs, with the majority being herbaceous perennials. Plants have indeterminate inflorescences, which are sometimes reduced to a single flower. The flowers have a short hypanthium and a single carpel with a short gynophore, and after fertilization produce fruits that are legumes.

Growth Habit

The Leguminosae have a wide variety of growth forms including trees, shrubs or herbaceous plants or even vines or lianas. The herbaceous plants can be annuals, biennials or perennials, without basal or terminal leaf aggregations. They are upright plants, epiphytes or vines. The latter support themselves by means of shoots that twist around a support or through cauline or foliar tendrils. Plants can be heliophytes, mesophytes or xerophytes.

Leaves

The leaves are usually alternate and compound. Most often they are even- or odd-pinnately compound (e.g. Caragana and Robinia respectively), often trifoliate (e.g. Trifolium, Medicago) and rarely palmately compound (e.g. Lupinus), in the Mimosoideae and the Caesalpinioideae commonly bipinnate (e.g. Acacia, Mimosa). They always have stipules, which can be leaf-like (e.g. Pisum), thorn-like (e.g. Robinia) or be rather inconspicuous. Leaf margins are entire or, occasionally, serrate. Both the leaves and the leaflets often have wrinkled pulvini to permit nastic movements. In some species, leaflets have evolved into tendrils (e.g. Vicia).

Many species have leaves with structures that attract ants that protect the plant from herbivore insects (a form of mutualism). Extrafloral nectaries are common among the Mimosoideae and the Caesalpinioideae, and are also found in some Faboideae (e.g. Vicia sativa). In some Acacia, the modified hollow stipules are inhabited by ants and are known as domatia.

Roots

Many Fabaceae host bacteria in their roots within structures called root nodules. These bacteria, known as rhizobia, have the ability to take nitrogen gas (N_2) out of the air and convert it to a form of nitrogen that is usable to the host plant (NO_3^- or NH_3). This process is called nitrogen fixation. The legume, acting as a host, and rhizobia, acting as a provider of usable nitrate, form a symbiotic relationship.

Flowers

The flowers often have five generally fused sepals and five free petals. They are generally hermaphrodite, and have a short hypanthium, usually cup shaped. There are normally ten stamens and one elongated superior ovary, with a curved style. They are usually arranged in indeterminate inflorescences. Fabaceae are typically entomophilous plants (i.e. they are pollinated by insects), and the flowers are usually showy to attract pollinators.

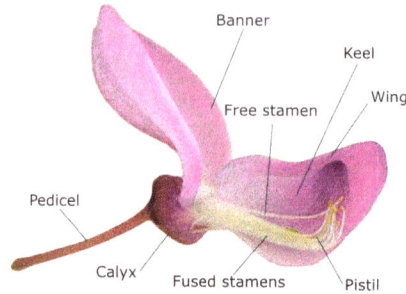

A flower of Wisteria sinensis, Faboideae. Two petals have been removed to show stamens and pistil

In the Caesalpinioideae, the flowers are often zygomorphic, as in Cercis, or nearly symmetrical with five equal petals in Bauhinia. The upper petal is the innermost one, unlike in the Faboideae. Some species, like some in the genus Senna, have asymmetric flowers, with one of the lower petals larger than the opposing one, and the style bent to one side. The calyx, corolla, or stamens can be showy in this group.

In the Mimosoideae, the flowers are actinomorphic and arranged in globose inflorescences. The petals are small and the stamens, which can be more than just 10, have long, coloured filaments, which are the showiest part of the flower. All of the flowers in an inflorescence open at once.

In the Faboideae, the flowers are zygomorphic, and have a specialized structure. The upper petal, called the banner, is large and envelops the rest of the petals in bud, often reflexing when the flower blooms. The two adjacent petals, the wings, surround the two bottom petals. The two bottom petals are fused together at the apex (remaining free at the base), forming a boat-like structure called the keel. The stamens are always ten in number, and their filaments can be fused in various configurations, often in a group of nine stamens plus one separate stamen. Various genes in the CYCLOIDEA (CYC)/ DICHOTOMA (DICH) family are expressed in the upper (also called dorsal or adaxial) petal; in some species, such as Cadia, these genes are expressed throughout the flower, producing a radially symmetrical flower.

Fruit

The ovary most typically develops into a legume. A legume is a simple dry fruit that usually dehisces (opens along a seam) on two sides. A common name for this type of fruit is a "pod", although that can also be applied to a few other fruit types. A few species have evolved samarae, loments, follicles, indehiscent legumes, achenes, drupes, and berries from the basic legume fruit.

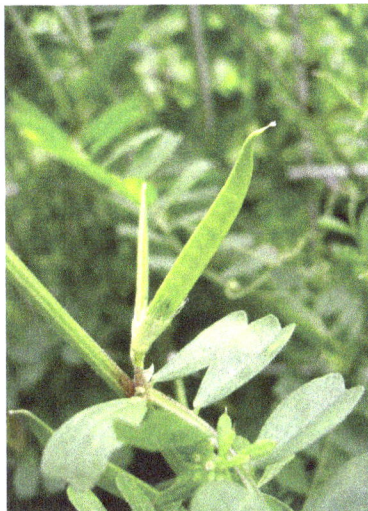

Legume of Vicia angustifolia

Physiology and Biochemistry

The Leguminosae are rarely cyanogenic, however, where they are, the cyanogenic compounds are derived from tyrosine, phenylalanine or leucine. They frequently contain alkaloids. Proanthocyanidins can be present either as cyanidin or delphinidine or both at the same time. Flavonoids such as kaempferol, quercitin and myricetin are often present. Ellagic acid has never been found in any of the genera or species analysed. Sugars are transported within the plants in the form of sucrose. C3 photosynthesis has been found in a wide variety of genera. The family has also evolved a unique chemistry. Pterocarpans are a class of molecules (derivatives of isoflavonoids) found only in the Fabaceae.

Ecology

Distribution and Habitat

The Fabaceae have an essentially worldwide distribution, being found everywhere except Antarctica and the high arctic. The trees are often found in tropical regions, while the herbaceous plants and shrubs are predominant outside the tropics.

Biological Nitrogen Fixation

Biological nitrogen fixation (BNF, performed by the organisms called diazotrophs) is a very old process that probably originated in the Archean eon when the primitive atmosphere lacked oxygen. It is only carried out by Euryarchaeota and just 6 of the more than 50 phyla of bacteria. Some of these lineages co-evolved together with the flowering plants establishing the molecular basis of a mutually beneficial symbiotic relationship. BNF is carried out in nodules that are mainly located in the root cortex, although they are occasionally located in the stem as in Sesbania rostrata. The spermatophytes that co-evolved with actinorhizal diazotrophs (Frankia) or with rhizobia to establish their symbiotic relationship belong to 11 families contained within the Rosidae clade (as established by the gene molecular phylogeny of rbcL, a gene coding for part of the RuBisCO enzyme in the chloroplast). This grouping indicates that the predisposition for forming nodules probably only arose once in flowering plants and that it can be considered as an ancestral characteristic that has been conserved or lost in certain lineages. However, such a wide distribution of families and genera within this lineage indicates that nodulation had multiple origins. Of the 10 families within the Rosidae, 8 have nodules formed by actinomyces (Betulaceae, Casuarinaceae, Coriariaceae, Datiscaceae, Elaeagnaceae, Myricaceae, Rhamnaceae and Rosaceae), and the two remaining families, Ulmaceae and Fabaceae have nodules formed by rhizobia.

Roots of Vicia with white root nodules visible.

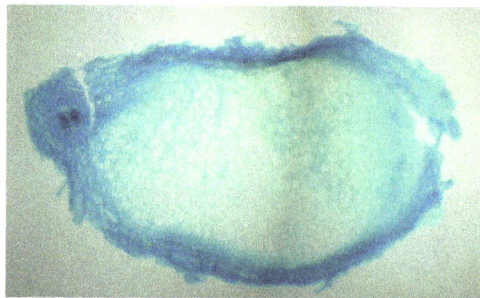

Cross-section through a root nodule of Vicia observed through a microscope.

The rhizobia and their hosts must be able to recognize each other for nodule formation to commence. Rhizobia are specific to particular host species although a rhizobia species may often infect more than one host species. This means that one plant species may be infected by more than one species of bacteria. For example, nodules in Acacia senegal can contain seven species of rhizobia belonging to three different genera. The most distinctive characteristics that allow rhizobia to be distinguished apart are the rapidity of their growth and the type of root nodule that they form with their host. Root nodules can be classified as being either indeterminate, cylindrical and often branched, and determinate, spherical with prominent lenticels. Indeterminate nodules are characteristic of legumes from temperate climates, while determinate nodules are commonly found in species from tropical or subtropical climates.

Nodule formation is common throughout the leguminosae, it is found in the majority of its members that only form an association with rhizobia, which in turn form an exclusive symbiosis with the leguminosae (with the exception of Parasponia, the only genus of the 18 Ulmaceae genera that is capable of forming nodules). Nodule formation is present in all the leguminosae sub-families, although it is less common in the Caesalpinioideae. All types of nodule formation are present in the sub-family Papilionoideae: indeterminate (with the meristem retained), determinate (without meristem) and the type included in Aeschynomene. The latter two are thought to be the most modern and specialised type of nodule as they are only present in some lines of the Papilionoideae sub-family. Even though nodule formation is common in the two monophyletic sub-families Papilionoideae and Mimosoideae they also contain species that do not form nodules. The presence or absence of nodule-forming species within the three sub-families indicates that nodule formation has arisen several times during the evolution of the leguminosae and that this ability has been lost in some lineages. For example, within the genus Acacia, a member of the Mimosoideae, A. pentagona does not form nodules, while other species of the same genus readily form nodules, as is the case for Acacia senegal, which forms both rapidly and slow growing rhizobial nodules.

Evolution, Phylogeny and Taxonomy

Evolution

The order Fabales contains around 7.3% of eudicot species and the greatest part of this diversity is contained in just one of the four families that order contains: Fabaceae. This clade also includes the Polygalaceae, Surianaceae and Quillajaceae families and its origins date back 94 to 89 million years, although it started its diversification some 79 to 74 million years ago. In fact, the Fabaceae have diversified during the early tertiary to become a ubiquitous part of the modern earth's biota, along with many other families belonging to the flowering plants.

The Fabaceae have an abundant and diverse fossil record, especially for the Tertiary period. Fossils of flowers, fruit, leaves, wood and pollen from this period have been found in numerous locations. The earliest fossils that can be definitively assigned to the Fabaceae appeared in the late Palaeocene (approximately 56 million years ago). Representatives of the 3 sub-families traditionally recognised as being members of the Fabaceae – Cesalpinioideae, Papilionoideae and Mimosoideae — as well as members of the large clades within these sub-families – such as the genistoides – have been found in periods a little later, starting between 55 and 50 million years ago. In fact, a wide variety of taxa representing the main lineages in the Fabaceae have been found in the fossil record dating from the middle to the late Eocene, suggesting that the majority of the modern Fabaceae groups were already present and that a broad diversification occurred during this period. Therefore, the Fabaceae started their diversification approximately 60 million years ago and the most important clades separated some 50 million years ago. The age of the main Cesalpinioideae clades have been estimated as between 56 and 34 million years and the basal group of the Mimosoideae as 44 ± 2.6 million years. The division between Mimosoideae and Faboideae is dated as occurring between 59 and 34 million years ago and the basal group of the Faboideae as 58.6 ± 0.2 million years ago. It has been possible to date the divergence of some of the groups within the Faboideae, even though diversification within each genus was relatively recent. For instance, Astragalus separated from the Oxytropis some 16 to 12 million years ago. In addition, the separation of the aneuploid species of Neoastragalus started 4 million years ago. Inga, another genus of the Papilionoideae with approximately 350 species, seems to have diverged in the last 2 million years.

It has been suggested, based on fossil and phylogenetic evidence, that legumes originally evolved in arid and/or semi-arid regions along the Tethys seaway during the Palaeogene Period. However, others contend that Africa (or even the Americas) cannot yet be ruled out as the origin of the family.

The current hypothesis about the evolution of the genes needed for nodulation is that they were recruited from other pathways after a polyploidy event. Several different pathways have been implicated as donating duplicated genes to the pathways need for nodulation. The main donors to the pathway were the genes associated with the arbuscular mycorrhiza symbiosis genes, the pollen tube formation genes and the haemoglobin genes. One of the main genes shown to be shared between the arbuscular mycorrhiza pathway and the nodulation pathway is SYMRK and it is involved in the plant-bacterial recognition. The pollen tube growth is similar to the infection thread development in that infection threads grow in a polar manner that is similar to a pollen tubes polar growth towards the ovules. Both pathways include the same type of enzymes, pectin-degrading cell wall enzymes. The enzymes needed to reduce nitrogen, nitrogenases, require a substantial input of ATP but at the same time are sensitive to free oxygen. To meet the requirements of this paradoxical situation, the plants express a type of haemoglobin called leghaemoglobin that is believed to be recruited after a duplication event. These three genetic pathways are believed to be part of a gene duplication event then recruited to work in nodulation.

Phylogeny and Taxonomy

Phylogeny

The phylogeny of the legumes has been the object of many studies by research groups from around the world. These studies have used morphology, DNA data (the chloroplast intron trnL, the chloroplast genes rbcL and matK, or the ribosomal spacers ITS) and cladistic analysis in order to investigate the relationships between the family's different lineages. The studies have confirmed that the traditional sub-families Mimosoideae and Papilionoideae are each monophyletic but both are nested within the paraphyletic sub-family Caesalpinioideae. All the different approaches have yielded similar results regarding the relationships between the family's main clades, as shown in the cladogram below.

Taxonomy

The Fabaceae are placed in the order Fabales according to most taxonomic systems, including the APG III system. The family includes three subfamilies:

- Mimosoideae: 80 genera and 3,200 species. Mostly tropical and warm temperate Asia and America. Mimosa, Acacia.

- Caesalpinioideae: 170 genera and 2,000 species, cosmopolitan. Caesalpinia, Senna, Bauhinia, Amherstia.

- Faboideae (Papilionoideae): 470 genera and 14,000 species, cosmopolitan. Astragalus, Lupinus.

These three subfamilies have been alternatively treated at the family level, as in the Cronquist and Dahlgren systems. However, this choice has not been supported by late 20th and early 21st century evidence, which has shown the Caesalpinioideae to be paraphyletic and the Fabaceae sensu lato to be monophyletic. While the Mimosoideae and the Faboideae are largely monophyletic, the Caesalpinioideae appear to be paraphyletic and the tribe Cercideae is probably sister to the rest of the family. Moreover, there are a number of genera whose placement into the Caesalpinioideae is not always agreed on (e.g. Dimorphandra).

Genera

The 730 genera included in this family can be viewed on the following three pages:

- List of Mimosoideae genera

- List of Caesalpinioideae genera

- List of Faboideae genera

Economic and Cultural Importance

Legumes are economically and culturally important plants due to their extraordinary diversity and abundance, the wide variety of edible vegetables they represent and due to the variety of uses they can be put to: in horticulture and agriculture, as a food, for the compounds they contain that have medicinal uses and for the oil and fats they contain that have a variety of uses.

Food and Forage

The history of legumes is tied in closely with that of human civilization, appearing early in Asia, the Americas (the common bean, several varieties) and Europe (broad beans) by 6,000 BCE, where they became a staple, essential as a source of protein.

Their ability to fix atmospheric nitrogen reduces fertilizer costs for farmers and gardeners who grow legumes, and means that legumes can be used in a crop rotation to replenish soil that has been depleted of nitrogen. Legume seeds and foliage have a comparatively higher protein content than non-legume materials, due to the additional nitrogen that legumes receive through the process. Some legume species perform hydraulic lift, which makes them ideal for intercropping.

Farmed legumes can belong to numerous classes, including forage, grain, blooms, pharmaceutical/industrial, fallow/green manure and timber species, with most commercially farmed species filling two or more roles simultaneously.

There are of two broad types of forage legumes. Some, like alfalfa, clover, vetch, and Arachis, are sown in pasture and grazed by livestock. Other forage legumes such as Leucaena or Albizia are woody shrub or tree species that are either broken down by livestock or regularly cut by humans to provide stock feed.

Grain legumes are cultivated for their seeds, and are also called pulses. The seeds are used for human and animal consumption or for the production of oils for industrial uses. Grain legumes include both herbaceous plants like beans, lentils, lupins, peas and peanuts. and trees such as carob, mesquite and tamarind.

Bloom legume species include species such as lupin, which are farmed commercially for their blooms as well as being popular in gardens worldwide. Laburnum, Robinia, Gleditsia, Acacia, Mimosa, and Delonix are ornamental trees and shrubs.

Industrial farmed legumes include Indigofera, cultivated for the production of indigo, Acacia, for gum arabic, and Derris, for the insecticide action of rotenone, a compound it produces.

Fallow or green manure legume species are cultivated to be tilled back into the soil to exploit the high nitrogen levels found in most legumes. Numerous legumes are farmed for this purpose, including Leucaena, Cyamopsis and Sesbania.

Various legume species are farmed for timber production worldwide, including numerous Acacia species, Dalbergia species, and Castanospermum australe.

Melliferous plants offer nectar to bees and other insects to encourage them to carry pollen from the flowers of one plant to others thereby ensuring pollination.A number of legume species are good nectar providers such as alfalfa, white clover, sweet clover and various Prosopis species. Many plants in the Fabaceae family are an important source of pollen for the bumblebee species Bombus hortorum. This bee species is especially fond of one species in particular; Trifolium pratense, also known as red clover, is a popular food source in the diet of Bombus hortorum.

Industrial uses

Natural Gums

Natural gums are vegetable exudates that are released as the result of damage to the plant such as that resulting from the attack of an insect or a natural or artificial cut. These exudates contain heterogeneous polysaccharides formed of different sugars and usually containing uronic acids. They form viscous colloidal solutions. There are different species that produce gums. The most important of these species belong to the leguminosae. They are widely used in the pharmaceutical, cosmetic, food and textile sectors. They also have interesting therapeutic properties; for example gum arabic is antitussive and anti-inflammatory. The most well known gums are tragacanth (Astragalus gummifer), gum arabic (Acacia senegal) and guar gum (Cyamopsis tetragonoloba).

Dyes

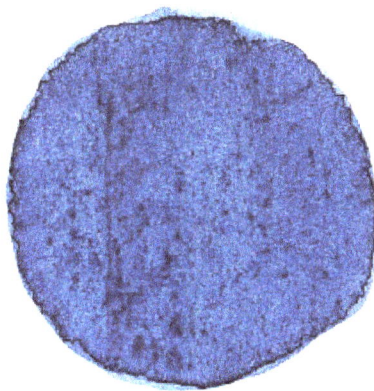

Indigo colorant

The species used to produce dyes include the following: Logwood Haematoxylon campechianum; a large spiny tree that can grow up to 15 m tall. Its cork is thin and soft and its wood is hard. The heartwood is used to produce dyes that are red and purple. The histological stain called haematoxylin is produced from this species. Brazilwood tree (Caesalpinia echinata) is similar to the previous tree but smaller and with red or purple flowers. The wood is also used to produce a red or purple dye. The Madras thorn (Pithecallobium dulce) is another spiny tree native to Latin America, it grows up to 4 m high and has yellow or green flowers that grow in florets. Its fruit is reddish and is used to produce a yellow dye. Indigo dye is extracted from the True indigo plant Indigofera tinctoria that is native to Asia. In Central and South America dyes are produced from two species related to this species, indigo from Indigofera suffruticosa and Natal indigo from Indigofera arrecta.yellow dye is extracted from Butea monosperma commonly called as flame of the forest.

Ornamentals

The Cockspur Coral Tree Erythrina crista-galli is one of many leguminosae used as ornamental plants. In addition, it is the National Flower of Argentina and Uruguay.

Legumes have been used as ornamental plants throughout the world for many centuries. Their vast diversity of heights, shapes, foliage and flower colour means that this family is commonly used in the design and planting of everything from small gardens to large parks. The following is a list of the main ornamental legume species, listed by sub-family.

- Subfamily Caesalpinioideae: Bauhinia forficata, Caesalpinia gilliesii, Caesalpinia spinosa, Ceratonia siliqua, Cercis siliquastrum, Gleditsia triacanthos, Gymnocladus dioica, Parkinsonia aculeata, Senna multiglandulosa.

- Subfamily Mimosoideae: Acacia caven, Acacia cultriformis, Acacia dealbata, Acacia karroo, Acacia longifolia, Acacia melanoxylon, Acacia paradoxa, Acacia retinodes, Acacia saligna, Acacia verticillata, Acacia visco, Albizzia julibrissin, Calliandra tweediei, Paraserianthes lophantha, Prosopis chilensis.

- Subfamily Faboideae: Clianthus puniceus, Citysus scoparius, Erythrina crista-galli, Erythrina falcata, Laburnum anagyroides, Lotus peliorhynchus, Lupinus arboreus, Lupinus polyphyllus, Otholobium glandulosum, Retama monosperma, Robinia hispida, Robinia luxurians, Robinia pseudoacacia, Sophora japonica, Sophora macnabiana, Sophora macrocarpa, Spartium junceum, Teline monspessulana, Tipuana tipu, Wisteria sinensis.

Emblematic Leguminosae

- The Cockspur Coral Tree (Erythrina crista-galli), is the National Flower of Argentina and Uruguay.

- The Elephant ear tree (Enterolobium cyclocarpum) is the national tree of Costa Rica, by Executive Order of 31 August 1959.

- The Brazilwood tree (Caesalpinia echinata) has been the national tree of Brazil since 1978.

- The Golden wattle Acacia pycnantha is Australia's national flower.

- The Hong Kong Orchid tree Bauhinia blakeana is the national flower of Hong Kong.

Brassicaceae

Brassicaceae or Cruciferae is a medium-sized and economically important family of flowering plants commonly known as the mustards, the crucifers, or the cabbage family.

The name Brassicaceae is derived from the included genus Brassica. The alternative older name, Cruciferae, meaning "cross-bearing", describes the four petals of mustard flowers, which resemble a cross. Cruciferae is one of eight plant family names without the suffix -aceae that are authorized alternative names (according to ICBN Art. 18.5 and 18.6 Vienna Code).

The family contains 372 genera and 4060 accepted species. The largest genera are Draba (440 species), Erysimum (261 species), Lepidium (234 species), Cardamine (233 species), and Alyssum (207 species).

The family contains the cruciferous vegetables, including species such as Brassica oleracea (e.g., broccoli, cabbage, cauliflower, kale, collards), Brassica rapa (turnip, Chinese cabbage, etc.), Brassica napus (rapeseed, etc.), Raphanus sativus (common radish), Armoracia rusticana (horseradish), Matthiola (stock) and the model organism Arabidopsis thaliana (thale cress).

Pieris rapae and other butterflies of the family Pieridae are some of the best-known pests of Brassicaceae species planted as commercial crops.

Taxonomy

The family is included in the Brassicales according to the APG system. Older systems (e.g., Arthur Cronquist's) placed them into the Capparales, a now-defunct order that had a similar definition.

This family comprises about 365 genera and 3200 species all over the world; 94 species of 38 genera are found in Nepal. The plants are mostly herbs. A close relationship has long been acknowledged between the Brassicaceae and the caper family, Capparaceae, in part because members of both groups produce glucosinolate (mustard oil) compounds. The Capparaceae as traditionally circumscribed were paraphyletic with respect to Brassicaceae, with Cleome and several related genera being more closely related to the Brassicaceae than to other Capparaceae. The APG II system, therefore, has merged the two families under the name Brassicaceae. Other classifications have continued to recognize the Capparaceae, but with a more restricted circumscription, either including Cleome and its relatives in the Brassicaceae or recognizing them in the segregate family Cleomaceae. The APG III system has recently adopted this last solution, but this may change as a consensus arises on this point. This article deals with Brassicaceae sensu stricto, i.e. treating the Cleomaceae and Capparaceae as segregated families.

Description

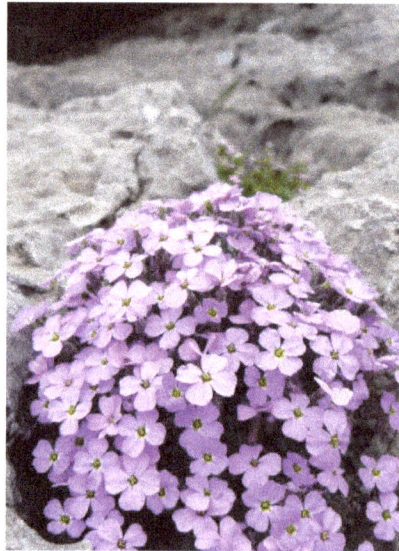

Aubrieta deltoidea (commonly known as purple rock cress) is a perennial wild flower used in gardening for its ornamental large inflorescence.

The family consists mostly of herbaceous plants with annual, biennial, or perennial lifespans. However, around the Mediterranean, they include also a dozen woody shrubs 1-3 m tall, e.g. in northern Africa (Zilla spinosa and Ptilotrichum spinosum), in the Dalmatian islands (Dendralyssum and Cramboxylon), and chiefly in Canarias with some woody cruciferous genera: Dendrosinapis, Descurainia, Parolinia, Stanleya, etc..

The leaves are alternate (rarely opposite), sometimes organized in basal rosettes; in rare shrubby crucifers of Mediterranean their leaves are mostly in terminal rosettes, and may be coriaceous and evergreen. They are very often pinnately incised and do not have stipules.

The structure of the flowers is extremely uniform throughout the family. They have four free saccate sepals and four clawed free petals, staggered. They can be disymmetric or slightly zygomorphic, with a typical cross-like arrangement (hence the name Cruciferae). They have six stamens, four of which are longer (as long as the petals) and are arranged in a cross like the petals and the other two are shorter (tetradynamous flower). The pistil is made up of two fused carpels and the style is very short, with two lobes. The ovary is superior. The flowers form ebracteate racemose inflorescences, often apically corymb-like.

Pollination occurs by entomogamy; nectar is produced at the base of the stamens and stored on the sepals.

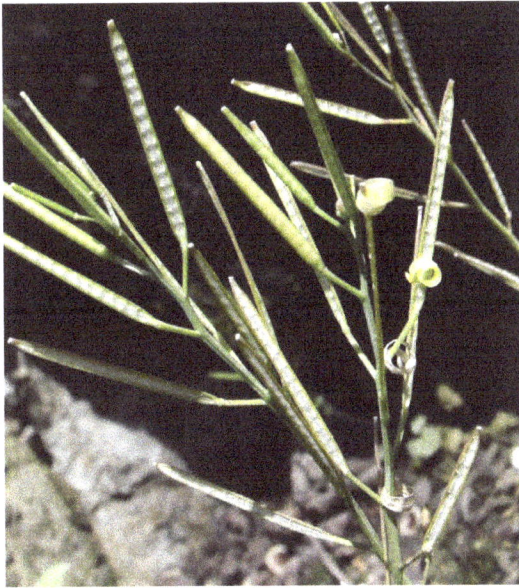

Siliquae of Cardamine impatiens

The fruit is a peculiar kind of capsule named siliqua (plural siliquae). It opens by two valves, which are the modified carpels, leaving the seeds attached to a framework made up of the placenta and tissue from the junction between the valves (replum). Often, an indehiscent beak occurs at the top of the style and one or more seeds may be borne there. Where a siliqua is less than three times as long as it is broad, it is usually termed a silicula. The siliqua may break apart at constrictions occurring between the segments of the seeds, thus forming a sort of loment (e.g., Raphanus), it may eject the seeds explosively (e.g., Cardamine) or may be evolved in a sort of samara (e.g., Isatis). The fruit is often the most important diagnostic character for plants in this family. Most members share a suite of glucosinolate compounds that have a typical pungent odour usually associated with cole crops.

Uses

The importance of this family for food crops has led to its selective breeding throughout history. Some examples of cruciferous food plants are the cabbage, broccoli, cauliflower, turnip, rapeseed, mustard, radish, horseradish, cress, wasabi, and watercress.

Lunaria annua with ripe seed pods

Smelowskia americana is endemic to the midlatitude mountains of western North America.

Cruciferous vegetables

Matthiola (stock), Cheiranthus, Lobularia, and Iberis (candytufts) are appreciated for their flowers. Lunaria (honesty) is cultivated for the decorative value of the translucent replum of the round silicula that remains on the dried stems after dehiscence.

Capsella bursa-pastoris, Lepidium, and many Cardamine species are common weeds.

Isatis tinctoria (woad) was used in the past to produce the colour indigo.

Arabidopsis thaliana is a very important model organism in the study of the flowering plants (Angiospermae).

Cucurbitaceae

The Cucurbitaceae, also called cucurbits and the gourd family, are a plant family consisting of ca 965 species in around 95 genera, the most important of which are:

- Cucurbita – squash, pumpkin, zucchini, some gourds

- Lagenaria – mostly inedible gourds

- Citrullus – watermelon (C. lanatus, C. colocynthis) and others

- Cucumis – cucumber (C. sativus), various melons

- Luffa – the common name is also luffa, sometimes spelled loofah (when fully ripened, two species of this fibrous fruit are the source of the loofah scrubbing sponge)

The plants in this family are grown around the tropics and in temperate areas, where those with edible fruits were among the earliest cultivated plants both in the Old and New Worlds. The Cucurbitaceae family ranks among the highest of plant families for number and percentage of species used as human food.

Pumpkins and squashes displayed in a show competition

The Cucurbitaceae consist of 98 proposed genera with 975 species, mainly in regions tropical and subtropical. All species are sensitive to frost. Most of the plants in this family are annual vines, but some are woody lianas, thorny shrubs, or trees (Dendrosicyos). Many species have large, yellow or white flowers. The stems are hairy and pentangular. Tendrils are present at 90° to the leaf petioles at nodes. Leaves are exstipulate alternate simple palmately lobed or palmately compound. The flowers are unisexual, with male and female flowers on different plants (dioecious) or on the same plant (monoecious). The female flowers have inferior ovaries. The fruit is often a kind of modified berry called a pepo.

Fossil History

One of the oldest fossil records so far is Cucurbitaciphyllum lobatum from the Paleocene epoch, found at Shirley Canal, Montana. It was described for the first time in 1924 by Knowlton. The fossil leaf is palmate, trilobed with rounded lobal sinuses and an entire or serrate margin. It has a leaf pattern similar to the members of the genera Kedrotis, Melothria and Zehneria.

Classification

A selection of cucurbits of the South Korean Genebank in Suwon

Cucurbits on display at the Real Jardín Botánico de Madrid, with the title "Variedades de calabaza"

The following is a classification given by Charles Jeffrey in 1990. However, a 2011 study based on genetics does not support this taxonomy with two subfamilies and eight tribes, but rather delineates 15 tribes, five of them new, consisting of 95 genera rather than Jeffrey's 121.

Subfamily Zanonioideae (small striate pollen grains)

- Tribe Zanonieae

    - Subtribe Fevilleinae: Fevillea

    - Subtribe Zanoniinae: Alsomitra Zanonia Siolmatra Gerrardanthus Zygosicyos Xerosicyos Neoalsomitra

    - Subtribe Gomphogyninae: Hemsleya Gomphogyne Gynostemma

- Subtribe Actinostemmatinae: Bolbostemma Actinostemma

- Subtribe Sicydiinae: Sicydium Chalema Pteropepon Pseudosicydium Cyclantheropsis

Subfamily Cucurbitoideae (styles united into a single column)

- Tribe Melothrieae

  - Subtribe Dendrosicyinae: Kedrostis Dendrosicyos Corallocarpus Ibervillea Tumamoca Halosicyos Ceratosanthes Doyerea Trochomeriopsis Seyrigia Dieterlea Cucurbitella Apodanthera Guraniopsis Melothrianthus Wilbrandia

  - Subtribe Guraniinae: Helmontia Psiguria Gurania

  - Subtribe Cucumerinae: Melancium Cucumeropsis Posadaea Melothria Muellarargia Zehneria Cucumis (including: Mukia, Dicaelospermum, Cucumella, Oreosyce, and Myrmecosicyos).

  - Subtribe Trochomeriinae: Solena Trochomeria Dactyliandra Ctenolepis

- Tribe Schizopeponeae: Schizopepon

- Tribe Joliffieae

  - Subtribe Thladianthinae: Indofevillea Siraitia Thladiantha Momordica

  - Subtribe Telfairiinae: Telfairia

- Tribe Trichosantheae

  - Subtribe Hodgsoniinae: Hodgsonia

  - Subtribe Ampelosicyinae: Ampelosicyos Peponium

  - Subtribe Trichosanthinae: Gymnopetalum Trichosanthes Tricyclandra

  - Subtribe Herpetosperminae: Cephalopentandra Biswarea Herpetospermum Edgaria

- Tribe Benincaseae

  - Subtribe Benincasinae: Cogniauxia Ruthalicia Lagenaria Benincasa Praecitrullus Citrullus Acanthosicyos Eureiandra Bambekea Nothoalsomitra Coccinia Diplocyclos Raphidiocystis Lemurosicyos Zombitsia Ecballium Bryonia

  - Subtribe Luffinae: Luffa

- Tribe Cucurbiteae (pantoporate, spiny pollen): Cucurbita Sicana Tecunumania

Calycophysum Peponopsis Anacaona Polyclathra Schizocarpum Penelopeia Cionosicyos Cayaponia Selysia Abobra

- Tribe Sicyeae (trichomatous nectary, four- to 10-colporate pollen grains)

    - Subtribe Cyclantherinae: Hanburia Echinopepon Marah Echinocystis Vaseyanthus Brandegea Apatzingania Cremastopus Elateriopsis Pseudocyclanthera Cyclanthera Rytidostylis

    - Subtribe Sicyinae: Sicyos Sicyosperma Parasicyos Microsechium Sechium Sechiopsis Pterosicyos

- Incertae sedis: Odosicyos

Alphabetical list of genera: Abobra Acanthosicyos Actinostemma Alsomitra Ampelosycios Anacaona Apatzingania Apodanthera Bambekea Benincasa Biswarea Bolbostemma Brandegea Bryonia Calycophysum Cayaponia Cephalopentandra Ceratosanthes Chalema Cionosicyos Citrullus Coccinia Cogniauxia Corallocarpus Cremastopus Ctenolepis Cucumella Cucumeropsis Cucumis Cucurbita Cucurbitella Cyclanthera Dactyliandra Dendrosicyos Dicaelospermum Dieterlea Diplocyclos Doyerea Ecballium Echinocystis Echinopepon Edgaria Elateriopsis Eureiandra Fevillea Gerrardanthus Gomphogyne Gurania Guraniopsis Gymnopetalum Gynostemma Halosicyos Hanburia Helmontia Hemsleya Herpetospermum Hodgsonia Ibervillea Indofevillea Kedrostis Lagenaria Lemurosicyos Luffa Marah Melancium Melothria Melothrianthus Microsechium Momordica Muellerargia Mukia Myrmecosicyos Neoalsomitra Nothoalsomitra Odosicyos Oreosyce Parasicyos Penelopeia Peponium Peponopsis Polyclathra Posadaea Praecitrullus Pseudocyclanthera Pseudosicydium Psiguria Pteropepon Pterosicyos Raphidiocystis Ruthalicia Rytidostylis Schizocarpum Schizopepon Sechiopsis Sechium Selysia Seyrigia Sicana Sicydium Sicyos Sicyosperma Siolmatra Siraitia Solena Tecunumania Telfairia Thladiantha Trichosanthes Tricyclandra Trochomeria Trochomeriopsis Tumacoca Vaseyanthus Wilbrandia Xerosicyos Zanonia Zehneria Zombitsia Zygosicyos

Round melons and elongate adzhur melons in Kursi church mosaic, Israel, near the Sea of Galilee

Images of Cucurbits in Byzantine Mosaics from Israel

Six cucurbit crops are represented in 23 Byzantine-era mosaics from Israel, these being round melons (Cucumis melo), watermelons (Citrullus lanatus), sponge gourds (Luffa aegyptiaca), snake melons (faqqous, Cucumis melo Flexuosus Group), adzhur melons (Cucumis melo Adzhur Group), and bottle gourds (Lagenaria siceraria). Cucurbits are represented in 23 of the 134 mosaics containing images of crop plants, a surprisingly high frequency of 17%. Several of the cucurbit images have not been found elsewhere, suggesting a diverse and highly developed local horticulture of cucurbits in Israel during the Byzantine era. Representations of mature sponge gourds are found in widespread localities, suggestive of the high value accorded to cleanliness and hygiene.

References

- Luther Burbank. Practical Orchard Plans and Methods: How to Begin and Carry on the Work. The Minerva Group. ISBN 1-4147-0141-1.

- Schlegel, Rolf H J (January 1, 2003). Encyclopedic Dictionary of Plant Breeding and Related Subjects. Haworth Press. p. 177. ISBN 1-56022-950-0.

- Mauseth, James D. (April 1, 2003). Botany: An Introduction to Plant Biology. Jones and Bartlett. pp. 271–272. ISBN 0-7637-2134-4.

- McGee, Harold (November 16, 2004). On Food and Cooking: The Science and Lore of the Kitchen. Simon & Schuster. pp. 247–248. ISBN 0-684-80001-2.

- Mauseth, James D. (2003). Botany: an introduction to plant biology. Boston: Jones and Bartlett Publishers. p. 258. ISBN 978-0-7637-2134-3.

- Rost, Thomas L.; Weier, T. Elliot; Weier, Thomas Elliot (1979). Botany: a brief introduction to plant biology. New York: Wiley. pp. 135–37. ISBN 0-471-02114-8.

- Singh, Gurcharan (2004). Plants Systematics: An Integrated Approach. Science Publishers. p. 83. ISBN 1-57808-351-6.

- Spiegel-Roy, P.; E. E. Goldschmidt (August 28, 1996). The Biology of Citrus. Cambridge University Press. pp. 87–88. ISBN 0-521-33321-0.

- Heiser, Charles B. (April 1, 2003). Weeds in My Garden: Observations on Some Misunderstood Plants. Timber Press. pp. 93–95. ISBN 0-88192-562-4.

- Farrell, Kenneth T. (November 1, 1999). Spices, Condiments and Seasonings. Springer. pp. 17–19. ISBN 0-8342-1337-0.

- Adams, Denise Wiles (February 1, 2004). Restoring American Gardens: An Encyclopedia of Heirloom Ornamental Plants, 1640–1940. Timber Press. ISBN 0-88192-619-1.

- Cothran, James R. (November 1, 2003). Gardens and Historic Plants of the Antebellum South. University of South Carolina Press. p. 221. ISBN 1-57003-501-6.

4

Hydroponics: An Integrated Study

Hydroponics is the method of growing plant without soil; instead of using soil, mineral nutrient solutions are used in a water solvent. Aquaponics, hydroponic dosers and historical hydroculture are some of the aspects of hydroponics that have been elucidated in the following section.

Hydroponics

NASA researcher checking hydroponic onions with Bibb lettuce to his left and radishes to the right

Hydroponics is a subset of hydroculture, the method of growing plants without soil, using mineral nutrient solutions in a water solvent. Terrestrial plants may be grown with only their roots exposed to the mineral solution, or the roots may be supported by an inert medium, such as perlite or gravel. The nutrients in hydroponics can be from fish waste, duck manure, or normal nutrients.

History

The earliest published work on growing terrestrial plants without soil was the 1627 book Sylva Sylvarum by Francis Bacon, printed a year after his death. Water culture became a popular research technique after that. In 1699, John Woodward published his water culture experiments with spearmint. He found that plants in less-pure water sources grew better than plants in distilled water. By 1842, a list of nine elements believed to be essential for plant growth had been compiled, and the discoveries of German botanists Julius von Sachs and Wilhelm Knop, in the years 1859–1875, resulted in a development of the technique of soilless cultivation. Growth of terrestrial plants without soil in mineral nutrient solutions was called solution culture. It quickly became a standard research and teaching technique and is still widely used. Solution culture is now considered a type of hydroponics where there is no inert medium.

In 1929, William Frederick Gericke of the University of California at Berkeley began publicly promoting that solution culture be used for agricultural crop production. He first termed it aquaculture but later found that aquaculture was already applied to culture of aquatic organisms. Gericke created a sensation by growing tomato vines twenty-five feet high in his back yard in mineral nutrient solutions rather than soil. He introduced the term hydroponics, water culture, in 1937, proposed to him by W. A. Setchell, a phycologist with an extensive education in the classics.

Reports of Gericke's work and his claims that hydroponics would revolutionize plant agriculture prompted a huge number of requests for further information. Gericke had been denied use of the University's greenhouses for his experiments due to the administration's skepticism, and when the University tried to compel him to release his preliminary nutrient recipes developed at home he requested greenhouse space and time to improve them using appropriate research facilities. While he was eventually provided greenhouse space, the University assigned Hoagland and Arnon to re-develop Gericke's formula and show it held no benefit over soil grown plant yields, a view held by Hoagland. In 1940, Gericke published the book, Complete Guide to Soil less Gardening, after leaving his academic position in a climate that was politically unfavorable.

Two other plant nutritionists at the University of California were asked to research Gericke's claims. Dennis R. Hoagland and Daniel I. Arnon wrote a classic 1938 agricultural bulletin, The Water Culture Method for Growing Plants Without Soil,. Hoagland and Arnon claimed that hydroponic crop yields were no better than crop yields with good-quality soils. Crop yields were ultimately limited by factors other than mineral nutrients, especially light. This research, however, overlooked the fact that hydroponics has other advantages including the fact that the roots of the plant have constant access to oxygen and that the plants have access to as much or as little water as they need. This is important as one of the most common errors when growing is over- and under- watering; and hydroponics prevents this from occurring as large amounts of water can be made available to the plant and any water not used, drained away, recirculated, or actively aerated, eliminating anoxic conditions, which drown root systems in soil. In soil, a grower needs to be very experienced to know exactly how much water to feed the plant. Too much and the plant will be unable to access oxygen; too little and the plant will lose the ability to transport nutrients, which are typically moved into the roots while in solution. These two researchers developed several formulas for mineral nutrient solutions, known as Hoagland solution. Modified Hoagland solutions are still in use.

One of the earliest successes of hydroponics occurred on Wake Island, a rocky atoll in the Pacific Ocean used as a refuelling stop for Pan American Airlines. Hydroponics was used there in the 1930s to grow vegetables for the passengers. Hydroponics was a necessity on Wake Island because there was no soil, and it was prohibitively expensive to airlift in fresh vegetables.

In the 1960s, Allen Cooper of England developed the Nutrient film technique. The Land Pavilion at Walt Disney World's EPCOT Center opened in 1982 and prominently features a variety of hydroponic techniques. In recent decades, NASA has done extensive hydroponic research for its Controlled Ecological Life Support System (CELSS). Hydroponics intended to take place on Mars are using LED lighting to grow in a different color spectrum with much less heat.

Plants that are not traditionally grown in a climate would be possible to grow using a controlled environment system like hydroponics. NASA has also looked to utilize hydroponics in the space program. Ray Wheeler, a plant physiologist at Kennedy Space Center's Space Life Science Lab, believes that hydroponics will create advances within space travel. He terms this as a bioregenerative life support system.

Techniques

There are two main variations for each medium, sub-irrigation and top irrigation[specify]. For all techniques, most hydroponic reservoirs are now built of plastic, but other materials have been used including concrete, glass, metal, vegetable solids, and wood. The containers should exclude light to prevent algae growth in the nutrient solution.

Static Solution Culture

The deep water raft tank at the CDC South Aquaponics greenhouse in Brooks, Alberta.

In static solution culture, plants are grown in containers of nutrient solution, such as glass Mason jars (typically, in-home applications), plastic buckets, tubs, or tanks. The solution is usually gently aerated but may be un-aerated. If un-aerated, the solution level is kept low enough that enough roots are above the solution so they get adequate oxygen. A hole is cut in the lid of the reservoir for each plant. There can be one to many plants per reservoir. Reservoir size can be increased as plant– size increases. A home made system can be constructed from plastic food containers or glass canning jars with aeration provided by an aquarium pump, aquarium airline tubing and aquarium valves. Clear containers are covered with aluminium foil, butcher paper, black plastic, or other material to exclude light, thus helping to eliminate the formation of algae. The nutrient solution is changed either on a schedule, such as once per week, or when the concentration drops below a certain level as determined with an electrical conductivity meter. Whenever the solution is depleted below a certain level, either water or fresh nutrient solution is added. A Mariotte's bottle, or a float valve, can be used to automatically maintain the solution level. In raft solution culture, plants are placed in a sheet of buoyant plastic that is floated on the surface of the nutrient solution. That way, the solution level never drops below the roots.

Continuous-Flow Solution Culture

The nutrient film technique being used to grow various salad greens

In continuous-flow solution culture, the nutrient solution constantly flows past the roots. It is much easier to automate than the static solution culture because sampling and adjustments to the temperature and nutrient concentrations can be made in a large storage tank that has potential to serve thousands of plants. A popular variation is the nutrient film technique or NFT, whereby a very shallow stream of water containing all the dissolved nutrients required for plant growth is recirculated past the bare roots of plants in a watertight thick root mat, which develops in the bottom of the channel and has an upper surface that, although moist, is in the air. Subsequent to this, an abundant supply of oxygen is provided to the roots of the plants. A properly designed NFT system is based on using the right channel slope, the right flow rate, and the right channel length. The main advantage of the NFT system over other forms of hydroponics is that the plant roots are exposed to adequate supplies of water, oxygen, and nutrients. In all other forms of production, there is a conflict between the supply of these requirements, since excessive or deficient amounts of one results in an imbalance of one or both of the others. NFT, because of its design, provides a system where all three requirements for healthy plant growth can be met at the same time, provided that the simple concept of

NFT is always remembered and practised. The result of these advantages is that higher yields of high-quality produce are obtained over an extended period of cropping. A downside of NFT is that it has very little buffering against interruptions in the flow (e.g. power outages). But, overall, it is probably one of the more productive techniques.

The same design characteristics apply to all conventional NFT systems. While slopes along channels of 1:100 have been recommended, in practice it is difficult to build a base for channels that is sufficiently true to enable nutrient films to flow without ponding in locally depressed areas. As a consequence, it is recommended that slopes of 1:30 to 1:40 are used. This allows for minor irregularities in the surface, but, even with these slopes, ponding and water logging may occur. The slope may be provided by the floor, or benches or racks may hold the channels and provide the required slope. Both methods are used and depend on local requirements, often determined by the site and crop requirements.

As a general guide, flow rates for each gully should be 1 liter per minute. At planting, rates may be half this and the upper limit of 2 L/min appears about the maximum. Flow rates beyond these extremes are often associated with nutritional problems. Depressed growth rates of many crops have been observed when channels exceed 12 metres in length. On rapidly growing crops, tests have indicated that, while oxygen levels remain adequate, nitrogen may be depleted over the length of the gully. As a consequence, channel length should not exceed 10–15 metres. In situations where this is not possible, the reductions in growth can be eliminated by placing another nutrient feed halfway along the gully and halving the flow rates through each outlet.

Aeroponics

Aeroponics is a system wherein roots are continuously or discontinuously kept in an environment saturated with fine drops (a mist or aerosol) of nutrient solution. The method requires no substrate and entails growing plants with their roots suspended in a deep air or growth chamber with the roots periodically wetted with a fine mist of atomized nutrients. Excellent aeration is the main advantage of aeroponics.

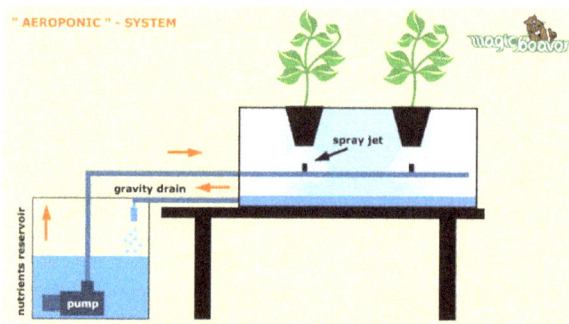

A diagram of the aeroponic technique.

Aeroponic techniques have proven to be commercially successful for propagation, seed germination, seed potato production, tomato production, leaf crops, and micro-greens.

Since inventor Richard Stoner commercialized aeroponic technology in 1983, aeroponics has been implemented as an alternative to water intensive hydroponic systems worldwide. The limitation of hydroponics is the fact that 1 kilogram (2.2 lb) of water can only hold 8 milligrams (0.12 gr) of air, no matter whether aerators are utilized or not.

Another distinct advantage of aeroponics over hydroponics is that any species of plants can be grown in a true aeroponic system because the micro environment of an aeroponic can be finely controlled. The limitation of hydroponics is that only certain species of plants can survive for so long in water before they become waterlogged. The advantage of aeroponics is that suspended aeroponic plants receive 100% of the available oxygen and carbon dioxide to the roots zone, stems, and leaves, thus accelerating biomass growth and reducing rooting times. NASA research has shown that aeroponically grown plants have an 80% increase in dry weight biomass (essential minerals) compared to hydroponically grown plants. Aeroponics used 65% less water than hydroponics. NASA also concluded that aeroponically grown plants requires ¼ the nutrient input compared to hydroponics. Unlike hydroponically grown plants, aeroponically grown plants will not suffer transplant shock when transplanted to soil, and offers growers the ability to reduce the spread of disease and pathogens. Aeroponics is also widely used in laboratory studies of plant physiology and plant pathology. Aeroponic techniques have been given special attention from NASA since a mist is easier to handle than a liquid in a zero-gravity environment.

Fogponics

Fogponics is a derivation of aeroponics wherein the nutrient solution is aerosolized by a diaphragm vibrating at ultrasonic frequencies. Solution droplets produced by this method tend to be 5-10 µm in diameter, smaller than those produced by forcing a nutrient solution through pressurized nozzles, as in aeroponics. The smaller size of the droplets allows them to diffuse through the air more easily, and deliver nutrients to the roots without limiting their access to oxygen.

Passive Sub-Irrigation

Passive sub-irrigation, also known as passive hydroponics or semi-hydroponics, is a method wherein plants are grown in an inert porous medium that transports water and fertilizer to the roots by capillary action from a separate reservoir as necessary, reducing labor and providing a constant supply of water to the roots. In the simplest method, the pot sits in a shallow solution of fertilizer and water or on a capillary mat saturated with nutrient solution. The various hydroponic media available, such as expanded clay and coconut husk, contain more air space than more traditional potting mixes, delivering increased oxygen to the roots, which is important in epiphytic plants such as orchids and bromeliads, whose roots are exposed to the air in nature. Additional advantages of passive hydroponics are the reduction of root rot and the additional ambient humidity provided through evaporations.

Ebb and Flow or Flood and Drain Sub-irrigation

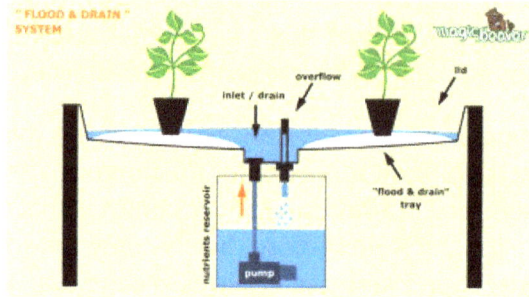

A Ebb and flow or flood and drain hydroponics system.

In its simplest form, there is a tray above a reservoir of nutrient solution. Either the tray is filled with growing medium (clay granules being the most common) and planted directly or pots of medium stand in the tray. At regular intervals, a simple timer causes a pump to fill the upper tray with nutrient solution, after which the solution drains back down into the reservoir. This keeps the medium regularly flushed with nutrients and air. Once the upper tray fills past the drain stop, it begins recirculating the water until the timer turns the pump off, and the water in the upper tray drains back into the reservoirs.

Run to Waste

In a run-to-waste system, nutrient and water solution is periodically applied to the medium surface. The method was invented in Bengal in 1946, for this reason it is sometimes referred to as "The Bengal System".

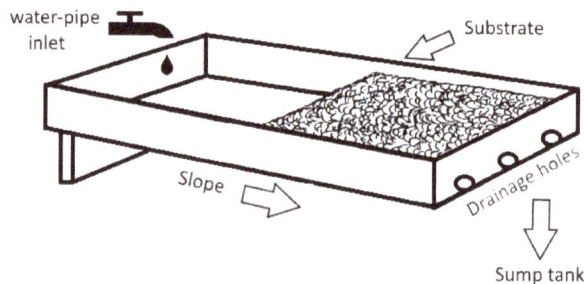

A run-to-waste hydroponics system referred to as "The Bengal System" after the region in northeastern India where it was invented (circa 1946–1948).

This method can be setup in various configurations. In its simplest form, a nutrient-and-water solution is manually applied one or more times per day to a container of inert growing media, such as rockwool, perlite, vermiculite, coco fibre, or sand. In a slightly more complex system, it is automated with a delivery pump, a timer and irrigation tubing to deliver nutrient solution with a delivery frequency that is governed by the key parameters of plant size, plant growing stage, climate, substrate, and substrate conductivity, pH, and water content.

In a commercial setting, watering frequency is multi-factorial and governed by computers or PLCs.

Commercial hydroponics production of large plants like tomatoes, cucumber, and peppers use one form or another of run-to-waste hydroponics.

In environmentally responsible uses, the nutrient rich waste is collected and processed through an on site filtration system to be used many times, making the system very productive.

The majority of bonsai are now grown in soil-free substrates (typically consisting of akadama, grit, diatomaceous earth and other inorganic components) and have their water and nutrients provided in a run-to-waste form.

Deep Water Culture

The Deep water culture technique being used to grow Hungarian wax peppers.

The hydroponic method of plant production by means of suspending the plant roots in a solution of nutrient-rich, oxygenated water. Traditional methods favor the use of plastic buckets and large containers with the plant contained in a net pot suspended from the centre of the lid and the roots suspended in the nutrient solution. The solution is oxygen saturated by an air pump combined with porous stones. With this method, the plants grow much faster because of the high amount of oxygen that the roots receive.

Top-Fed Deep Water Culture

Top-fed deep water culture is a technique involving delivering highly oxygenated nutrient solution direct to the root zone of plants. While deep water culture involves the plant roots hanging down into a reservoir of nutrient solution, in top-fed deep water culture the solution is pumped from the reservoir up to the roots (top feeding). The water is released over the plant's roots and then runs back into the reservoir below in a constantly recirculating system. As with deep water culture, there is an airstone in the reservoir that pumps air into the water via a hose from outside the reservoir. The airstone helps add oxygen to the water. Both the airstone and the water pump run 24 hours a day.

The biggest advantage of top-fed deep water culture over standard deep water culture is increased growth during the first few weeks. With deep water culture, there is a time when the roots have not reached the water yet. With top-fed deep water culture, the roots get easy access to water from the beginning and will grow to the reservoir below much more quickly than with a deep water culture system. Once the roots have reached the reservoir below, there is not a huge advantage with top-fed deep water culture over standard deep water culture. However, due to the quicker growth in the beginning, grow time can be reduced by a few weeks.

Rotary

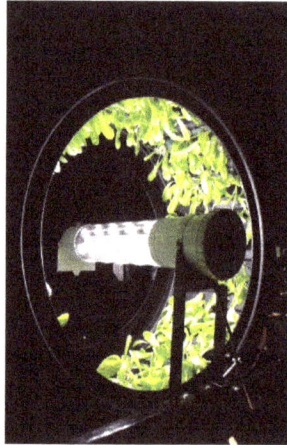

A Rotary hydroponic cultivation demonstration at the Belgian Pavilion Expo in 2015.

A rotary hydroponic garden is a style of commercial hydroponics created within a circular frame which rotates continuously during the entire growth cycle of whatever plant is being grown.

While system specifics vary, systems typically rotate once per hour, giving a plant 24 full turns within the circle each 24-hour period. Within the center of each rotary hydroponic garden is a high intensity grow light, designed to simulate sunlight, often with the assistance of a mechanized timer.

Each day, as the plants rotate, they are periodically watered with a hydroponic growth solution to provide all nutrients necessary for robust growth. Due to the plants continuous fight against gravity, plants typically mature much more quickly than when grown in soil or other traditional hydroponic growing systems. Due to the small foot print a rotary hydroponic system has, it allows for more plant material to be grown per square foot of floor space than other traditional hydroponic systems.

Substrates

One of the most obvious decisions hydroponic farmers have to make is which medium they should use. Different media are appropriate for different growing techniques.

Expanded Clay Aggregate

Expanded clay pebbles.

Baked clay pellets are suitable for hydroponic systems in which all nutrients are carefully controlled in water solution. The clay pellets are inert, pH neutral and do not contain any nutrient value.

The clay is formed into round pellets and fired in rotary kilns at 1,200 °C (2,190 °F). This causes the clay to expand, like popcorn, and become porous. It is light in weight, and does not compact over time. The shape of an individual pellet can be irregular or uniform depending on brand and manufacturing process. The manufacturers consider expanded clay to be an ecologically sustainable and re-usable growing medium because of its ability to be cleaned and sterilized, typically by washing in solutions of white vinegar, chlorine bleach, or hydrogen peroxide (H_2O_2), and rinsing completely.

Another view is that clay pebbles are best not re-used even when they are cleaned, due to root growth that may enter the medium. Breaking open a clay pebble after a crop has been shown to reveal this growth.

Growstones

Growstones, made from glass waste, have both more air and water retention space than perlite and peat. This aggregate holds more water than parboiled rice hulls. Growstones by volume consist of 0.5 to 5% calcium carbonate – for a standard 5.1 kg bag of Growstones that corresponds to 25.8 to 258 grams of calcium carbonate. The remainder is soda-lime glass.

Coir Peat

Coco peat, also known as coir or coco, is the leftover material after the fibres have been removed from the outermost shell (bolster) of the coconut. Coir is a 100% natural grow and flowering medium. Coconut coir is colonized with trichoderma fungi, which protects roots and stimulates root growth. It is extremely difficult to over-water coir due to its perfect air-to-water ratio; plant roots thrive in this environment. Coir has a high cation exchange, meaning it can store unused minerals to be released to the plant as and when it requires it. Coir is available in many forms; most common is coco peat, which has the appearance and texture of soil but contains no mineral content.

Rice Husks

Rice husks, a hydroponic growing substrate option.

Parboiled rice husks (PBH) are an agricultural byproduct that would otherwise have little use. They decay over time, and allow drainage, and even retain less water than growstones. A study showed that rice husks did not affect the effects of plant growth regulators.

Perlite

Perlite, a hydroponic growing substrate option.

Perlite is a volcanic rock that has been superheated into very lightweight expanded glass pebbles. It is used loose or in plastic sleeves immersed in the water. It is also used in potting soil mixes to decrease soil density. Perlite has similar properties and uses to vermiculite but, in general, holds more air and less water. If not contained, it can float if flood and drain feeding is used. It is a fusion of granite, obsidian, pumice and basalt. This volcanic rock is naturally fused at high temperatures undergoing what is called "Fusionic Metamorphosis".

Vermiculite

Vermiculite close-up.

Like perlite, vermiculite is a mineral that has been superheated until it has expanded into light pebbles. Vermiculite holds more water than perlite and has a natural "wicking" property that can draw water and nutrients in a passive hydroponic system. If too much water and not enough air surrounds the plants roots, it is possible to gradually lower the medium's water-retention capability by mixing in increasing quantities of perlite.

Pumice

A pumice stone.

Like perlite, pumice is a lightweight, mined volcanic rock that finds application in hydroponics.

Sand

Sand is cheap and easily available. However, it is heavy, does not hold water very well, and it must be sterilized between uses.

Gravel

The same type that is used in aquariums, though any small gravel can be used, provided it is washed first. Indeed, plants growing in a typical traditional gravel filter bed, with water circulated using electric powerhead pumps, are in effect being grown using gravel hydroponics. Gravel is inexpensive, easy to keep clean, drains well and will not become waterlogged. However, it is also heavy, and, if the system does not provide continuous water, the plant roots may dry out.

Wood Fibre

Excelsior, or wood wool

Wood fibre, produced from steam friction of wood, is a very efficient organic substrate for hydroponics. It has the advantage that it keeps its structure for a very long time. Wood wool (i.e. wood slivers) have been used since the earliest days of the hydroponics research. However, more recent research suggests that wood fibre may have detrimental effects on "plant growth regulators".

Sheep Wool

Wool from shearing sheep is a little-used yet promising renewable growing medium. In a study comparing wool with peat slabs, coconut fibre slabs, perlite and rockwool slabs to grow cucumber plants, sheep wool had a greater air capacity of 70%, which decreased with use to a comparable 43%, and water capacity that increased from 23% to 44% with use. Using sheep wool resulted in the greatest yield out of the tested substrates, while application of a biostimulator consisting of humic acid, lactic acid and Bacillus subtilis improved yields in all substrates.

Rock Wool

Rock wool (mineral wool) is the most widely used medium in hydroponics. Rock wool is an inert substrate suitable for both run-to-waste and recirculating systems. Rock wool is made from molten rock, basalt or 'slag' that is spun into bundles of single filament fibres, and bonded into a medium capable of capillary action, and is, in effect, protected from most common microbiological degradation. Rock wool has many advantages and some disadvantages. The latter being the possible skin irritancy (mechanical) whilst handling (1:1000). Flushing with cold water usually brings relief. Advantages include its proven efficiency and effectiveness as a commercial hydroponic substrate. Most of the rock wool sold to date is a non-hazardous, non-carcinogenic material, falling under Note Q of the European Union Classification Packaging and Labeling Regulation (CLP).

Rock wool close-up.

Brick Shards

Brick shards have similar properties to gravel. They have the added disadvantages of possibly altering the pH and requiring extra cleaning before reuse.

Polystyrene Packing Peanuts

Polystyrene foam peanuts

Polystyrene packing peanuts are inexpensive, readily available, and have excellent drainage. However, they can be too lightweight for some uses. They are used mainly in closed-tube systems. Note that polystyrene peanuts must be used; biodegradable packing peanuts will decompose into a sludge. Plants may absorb styrene and pass it to their consumers; this is a possible health risk.

Nutrient Solutions

Inorganic Hydroponic Solutions

The formulation of hydroponic solutions is an application of plant nutrition, with nutrient deficiency symptoms mirroring those found in traditional soil based agriculture.

However, the underlying chemistry of hydroponic solutions can differ from soil chemistry in many significant ways. Important differences include:

- Unlike soil, hydroponic nutrient solutions do not have cation-exchange capacity (CEC) from clay particles or organic matter. The absence of CEC means the pH and nutrient concentrations can change much more rapidly in hydroponic setups than is possible in soil.

- Selective absorption of nutrients by plants often imbalances the amount of counterions in solution. This imbalance can rapidly affect solution pH and the ability of plants to absorb nutrients of similar ionic charge. For instance, nitrate anions are often consumed rapidly by plants to form proteins, leaving an excess of cations in solution. This cation imbalance can lead to deficiency symptoms in other cation based nutrients (e.g. Mg^{2+}) even when an ideal quantity of those nutrients are dissolved in the solution.

- Depending the on pH, and/or the presence of water contaminants, nutrients, such as iron, can precipitate from the solution and become unavailable to plants. Routine adjustments to pH, buffering the solution, and/or the use of chelating agents is often necessary.

As in conventional agriculture, nutrients should be adjusted to satisfy Liebig's law of the minimum for each specific plant variety. Nevertheless, generally acceptable concentrations for nutrient solutions exist, with minimum and maximum concentration ranges for most plants being somewhat similar. Most nutrient solutions are mixed to have concentrations between 1,000 and 2,500 ppm. Acceptable concentrations for the individual nutrient ions, which comprise that total ppm figure, are summarized in the following table. For essential nutrients, concentrations below these ranges often lead to nutrient deficiencies while exceeding these ranges can lead to nutrient toxicity. Optimum nutrition concentrations for plant varieties are found empirically by experience and/or by plant tissue tests.

Element	Role	Ionic form(s)	Low range (ppm)	High range (ppm)	Common Sources	Comment
Nitrogen	Essential macronutrient	NO_3^- and/or NH_4^+	100	1000	KNO_3, NH_4NO_3, $Ca(NO_3)_2$, HNO_3, $(NH_4)_2SO_4$, and $(NH_4)_2HPO_4$	NH_4^+ interferes with Ca^{2+} uptake and can be toxic to plants if used as a major nitrogen source. A 3:1 ratio of NO_3^- to NH_4^+ is sometimes recommended to balance pH during nitrogen absorption.

Potassium	Essential macronutrient	K^+	100	400	KNO_3, K_2SO_4, KCl, KOH, K_2CO_3, K_2HPO_4, and K_2SiO_3	High concentrations interfere with the function Fe, Mn, and Zn. Zinc deficiencies often are the most apparent.
Phosphorus	Essential macronutrient	PO_4^{3-}	30	100	K_2HPO_4, KH_2PO_4, $NH_4H_2PO_4$, H_3PO_4, and $Ca(H_2PO_4)_2$	Excess NO_3^- tends to inhibit PO_4^{3-} absorption. The ratio of iron to PO_4^{3-} can affect co-preciptiation reactions.
Calcium	Essential macronutrient	Ca^{2+}	200	500	$Ca(NO_3)_2$, $Ca(H_2PO_4)$, $CaSO_4$, $CaCl_2$	Excess Ca^{2+} inhibits Mg^{2+} uptake.
Magnesium	Essential macronutrient	Mg^{2+}	50	100	$MgSO_4$ and $MgCl_2$	Should not exceed Ca^{2+} concentration due to competitive uptake.
Sulfur	Essential macronutrient	SO_4^{2-}	50	1000	$MgSO_4$, K_2SO_4, $CaSO_4$, H_2SO_4, $(NH_4)_2SO_4$, $ZnSO_4$, $CuSO_4$, $FeSO_4$, and $MnSO_4$	Unlike most nutrients, plants can tolerate a high concentration of the SO_4^{2-}, selectively absorbing the nutrient as needed. Undesirable counterion affects still apply however.
Iron	Essential micronutrient	Fe^{3+} and Fe^{2+}	2	5	FeDTPA, FeEDTA, iron citrate, iron tartrate, $FeCl_3$, and $FeSO_4$	pH values above 6.5 greatly decreases iron solubility. Chelating agents (e.g. DTPA, citric acid, or EDTA) are often added to increase iron solubility over a greater pH range.
Zinc	Essential micronutrient	Zn^{2+}	0.05	1	$ZnSO_4$	Excess zinc is highly toxic to plants but is essential for plants at low concentrations.
Copper	Essential micronutrient	Cu^{2+}	0.01	1	$CuSO_4$	Plant sensitivity to copper is highly variable. 0.1 ppm can cause toxic to some plants while a concentration up to 0.5 ppm for many plants is often considered ideal.
Manganese	Essential micronutrient	Mn^{2+}	0.5	1	$MnSO_4$ and $MnCl_2$	Uptake is enhanced by high PO_4^{3-} concentrations.
Boron	Essential micronutrient	$B(OH)_4^-$	0.3	10	H_3BO_3, and $Na_2B_4O_7$	An essential nutrient, however, some plants are highly sensitive to boron (e.g. toxic affects are apparent in citrus trees at 0.5 ppm).

Molybdenum	Essential micronutrient	MoO_4^-	0.001	0.05	$(NH_4)_6Mo_7O_{24}$ and Na_2MoO_4	A component of the enzyme nitrate reductase and required by rhizobia for nitrogen fixation.
Nickel	Essential micronutrient	Ni^{2+}	0.057	1.5	$NiSO_4$ and $NiCO_3$	Essential to many plants (e.g. legumes and some grain crops). Also used in the enzyme urease.
Chlorine	Variable micronutrient	Cl^-	0	Highly variable	KCl, $CaCl_2$, $MgCl_2$, and $NaCl$	Can interfere with NO_3^- uptake in some plants but can be beneficial in some plants (e.g. in asparagus at 5 ppm). Absent in conifers, ferns, and most bryophytes.
Aluminum	Variable micronutrient	Al^{3+}	0	10	$Al_2(SO_4)_3$	Essential for some plants (e.g. peas, maize, sunflowers, and cereals). Can be toxic to some plants below 10 ppm. Sometimes used to produce flower pigments (e.g. by Hydrangeas).
Silicon	Variable micronutrient	SiO_3^{2-}	0	140	K_2SiO_3, Na_2SiO_3, and H_2SiO_3	Present in most plants, abundant in cereal crops, grasses, and tree bark. Evidence that SiO_3^{2-} improves plant disease resistance exists.
Titanium	Variable micronutrient	Ti^{3+}	0	5	H_4TiO_4	Might be essential but trace Ti^{3+} is so ubiquitous that its addition is rarely warranted. At 5 ppm favorable growth effects in some crops are notable (e.g. pineapple and peas).
Cobalt	Non-essential micronutrient	Co^{2+}	0	0.1	$CoSO_4$	Required by rhizobia, important for legume root nodulation.
Sodium	Non-essential micronutrient	Na^+	0	Highly variable	Na_2SiO_3, Na_2SO_4, $NaCl$, $NaHCO_3$, and $NaOH$	Na^+ can partially replace K^+ in some plant functions but K^+ is still an essential nutrient.
Vanadium	Non-essential micronutrient	VO^{2+}	0	Trace, undetermined	$VOSO_4$	Beneficial for rhizobial N_2 fixation.
Lithium	Non-essential micronutrient	Li^+	0	Undetermined	Li_2SO_4, $LiCl$, and $LiOH$	Li^+ can increase the chlorophyll content of some plants (e.g. potato and pepper plants).

Organic Hydroponic Solutions

Organic fertilizers can be used to supplement or entirely replace the inorganic compounds used in conventional hydroponic solutions. However, using organic fertilizers introduces a number of challenges that are not easily resolved. Examples include:

- organic fertilizers are highly variable in their nutritional compositions. Even similar materials can differ significantly based on their source (e.g. the quality of manure varies based on an animal's diet).

- organic fertilizers are often sourced from animal byproducts, making disease transmission a serious concern for plants grown for human consumption or animal forage.

- organic fertilizers are often particulate and can clog substrates or other growing equipment. Sieving and/or milling the organic materials to fine dusts is often necessary.

- some organic materials (i.e. particularly manures and offal) can further degrade to emit foul odors.

Nevertheless, if precautions are taken, organic fertilizers can be used successfully in hydroponics.

Organically Sourced Macronutrients

Examples of suitable materials, with their average nutritional contents tabulated in terms of percent dried mass, are listed in the following table.

Organic material	N	P_2O_5	K_2O	CaO	MgO	SO_2	Comment
Bloodmeal	13.0%	2.0%	1.0%	0.5%	–	–	
Bone ashes	–	35.0%	–	46.0%	1.0%	0.5%	
Bonemeal	4.0%	22.5%	–	33.0%	0.5%	0.5%	
Hoof / Horn meal	14.0%	1.0%	–	2.5%	–	2.0%	
Fishmeal	9.5%	7.0%	–	0.5%	–	–	
Wool waste	3.5%	0.5%	2.0%	0.5%	–	–	
Wood ashes	–	2.0%	5.0%	33.0%	3.5%	1.0%	
Cottonseed ashes	–	5.5%	27.0%	9.5%	5.0%	2.5%	
Cottonseed meal	7.0%	3.0%	2.0%	0.5%	0.5%	–	

Dried locust or grasshopper	10.0%	1.5%	0.5%	0.5%	–	–	
Leather waste	5.5% to 22%	–	–	–	–	–	Milled to a fine dust.
Kelp meal, liquid seaweed	1%	–	12%	–	–	–	Commercial products available.
Poultry manure	2% to 5%	2.5% to 3%	1.3% to 3%	4.0%	1.0%	2.0%	A liquid compost which is sieved to remove solids and checked for pathogens.
Sheep manure	2.0%	1.5%	3.0%	4.0%	2.0%	1.5%	Same as poultry manure.
Goat manure	1.5%	1.5%	3.0%	2.0%	–	–	Same as poultry manure.
Horse manure	3% to 6%	1.5%	2% to 5%	1.5%	1.0%	0.5%	Same as poultry manure.
Cow manure	2.0%	1.5%	2.0%	4.0%	1.1%	0.5%	Same as poultry manure.
Bat guano	8.0%	40%	29%	Trace	Trace	Trace	High in micronutrients. Commercially available.
Bird guano	13%	8%	20%	Trace	Trace	Trace	High in micronutrients. Commercially available.

Organically Sourced Micronutrients

Micronutrients can be sourced from organic fertilizers as well. For example, composted pine bark is high in manganese and is sometimes used to fulfill that mineral requirement in hydroponic solutions. To satisfy requirements for National Organic Programs, pulverized, unrefined minerals (e.g. Gypsum, Calcite, and glauconite) can also be added to satisfy a plant's nutritional needs.

Additives

In addition to chelating agents, humic acids can be added to increase nutrient uptake.

Tools

Common Equipment

Managing nutrient concentrations and pH values within acceptable ranges is essential for successful hydroponic horticulture. Common tools used to manage hydroponic solutions include:

- Electrical conductivity meters, a tool which estimates nutrient ppm by measuring how well a solution transmits an electric current.

- pH meter, a tool that uses an electric current to determine the concentration of hydrogen ions in solution.

- Litmus paper, disposable pH indicator strips that determine hydrogen ion concentrations by color changing chemical reaction.

- Graduated cylinders or measuring spoons to measure out premixed, commercial hydroponic solutions.

Advanced Equipment

Advanced equipment can also be used to perform accurate chemical analyses of nutrient solutions. Examples include:

- Balances for accurately measuring materials.

- Laboratory glassware, such as burettes and pipettes, for performing titrations.

- Colorimeters for solution tests which apply the Beer–Lambert law.

Using advanced equipment for hydroponic solutions can be beneficial to growers of any background because nutrient solutions are often reusable. Because nutrient solutions are virtually never completely depleted, and should never be due to the unacceptably low osmotic pressure that would result, re-fortification of old solutions with new nutrients can save growers money and can control point source pollution, a common source for the eutrophication of nearby lakes and streams.

Software

Although pre-mixed concentrated nutrient solutions are generally purchased from commercial nutrient manufacturers by hydroponic hobbyists and small commercial growers, several tools exist to help anyone prepare their own solutions without extensive knowledge about chemistry. The free and open source tools HydroBuddy and HydroCal have been created by professional chemists to help any hydroponics grower prepare their own nutrient solutions. The first program is available for Windows, Mac and Linux while the second one can be used through a simple JavaScript interface. Both programs allow for basic nutrient solution preparation although HydroBuddy provides added functionality to use and save custom substances, save formulations and predict electrical conductivity values.

Mixing Solutions

Often mixing hydroponic solutions using individual salts is impractical for hobbyists and/or small-scale commercial growers because commercial products are available at reasonable prices. However, even when buying commercial products, multi-component fertilizers are popular. Often these products are bought as three part formulas which emphasize certain nutritional roles. For example, solutions for vegetative growth (i.e. high in nitrogen), flowering (i.e. high in potassium and phosphorus), and micro-nutrient solutions (i.e. with trace minerals) are popular. The timing and application of these multi-part fertilizers should coincide with a plant's growth stage. For example, at the end of an annual plant's life cycle, a plant should be restricted from high nitrogen fertilizers. In most plants, nitrogen restriction inhibits vegetative growth and helps induce flowering.

Advancements

With pest problems reduced and nutrients constantly fed to the roots, productivity in hydroponics is high; however, growers can further increase yield by manipulating a plant's environment by constructing sophisticated growrooms.

CO_2 Enrichment

To increase yield further, some sealed greenhouses inject CO_2 into their environment to help improve growth and plant fertility.

Organic Hydroponics

Organic hydroponics is a hydroponics culture system which is managed based on organic agriculture concepts. Most studies have focused on use of organic fertilizer. Conventional hydroponics cannot use organic fertilizers because organic compounds contained in hydroponic solution inhibit the growth of the crop roots, so it uses only inorganic fertilizer.

In this method of organic hydroponics, organic fertilizer is degraded into inorganic nutrients by microorganisms in the hydroponic solution via ammonification and nitrification. The microorganisms are cultured with a method of multiple parallel mineralization. The culture solution can be used as the hydroponic solution. Practical method of organic hydroponics is developed in National Agriculture and Food Research Organization (NARO), in Japan, in 2005.

History of Organic Hydroponics

Studies for establishing organic hydroponics have been conducted for a long time. Kennedy Space Center had studied organic hydroponics for crop production in space. It was necessary to develop the method to generate nitrate from organic fertilizer via ammonification and nitrification, because most of crops are nitrate-phylic but not ammonium-philic. It is easy to generate ammonium from organic fertilizer by saprophytic microorganisms. However it was difficult to degrade organic fertilizer to nitrate efficiently because the growth of nitrifying bacteria, such as the obligate chemolithoautotrophs Nitrosomonas spp. and Nitrospira spp., is particularly inhibited by the presence of organic compounds (Jensen 1950; Quastel and Scholefield 1951; Rittenberg 1969; Smith and Hoare 1977; Krummel and Harms 1982; Takahashi et al. 1992; Stutte 1996; Xu et al. 2000; Tomiyama et al. 2001).

Shinohara invented the method to efficiently generate nitrate from in water. The method, multiple parallel mineralization, consists of three manipulations: small inoculation of soil microorganisms, addition of small amounts of organic fertilizer, and aeration. The mineralized solution can be used as the hydroponic solution and organic fertilizer can be added in the solution during cultivation. This is the first practical organic hydroponics technique that organic fertilizer can be added directly during cultivation.

Farajollahzade et al., introduced a method in which soil remained as the source of nutrients and the nitrogen was supplied by a in-line tank containing nitrogen-fixing bacteria. This method could be considered as a variation of soil culture in which the plant and soil are connected by cycling liquid. By use of organic soil as the source of nutrients, they believe that it could be considered within the definition of organic agriculture.

Aquaponics

A small, portable aquaponics system. The term aquaponics is a portmanteau of the terms aquaculture and hydroponic agriculture.

Aquaponics refers to any system that combines conventional aquaculture (raising aquatic animals such as snails, fish, crayfish or prawns in tanks) with hydroponics (cultivating plants in water) in a symbiotic environment. In normal aquaculture, excretions from the animals being raised can accumulate in the water, increasing toxicity. In an aquaponic system, water from an aquaculture system is fed to a hydroponic system where the by-products are broken down by Nitrifying bacteria initially into nitrites and subsequently into nitrates , which are utilized by the plants as nutrients, and the water is then recirculated back to the aquaculture system.

As existing hydroponic and aquaculture farming techniques form the basis for all aquaponics systems, the size, complexity, and types of foods grown in an aquaponics system can vary as much as any system found in either distinct farming discipline.

History

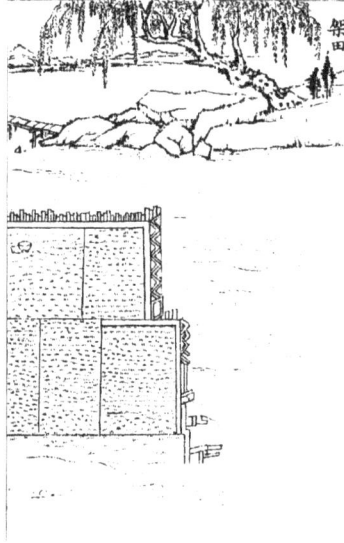

Woodcut from the 13th century Chinese agricultural manual Wang Zhen's Book on Farming (王禎農書) showing rice grown in a floating raft planter system in a pond

Aquaponics has ancient roots, although there is some debate on its first occurrence:

- Aztec cultivated agricultural islands known as chinampas in a system considered by some to be the first form of aquaponics for agricultural use where plants were raised on stationary (and sometime movable) islands in lake shallows and waste materials dredged from the Chinampa canals and surrounding cities were used to manually irrigate the plants.

- South China, Thailand, and Indonesia who cultivated and farmed rice in paddy fields in combination with fish are cited as examples of early aquaponics systems. These polycultural farming systems existed in many Far Eastern countries and raised fish such as the oriental loach (泥鰍, ドジョウ), swamp eel (黄鱔, 田鰻), common carp (鯉魚, コイ) and crucian carp (鯽魚) as well as pond snails (田螺) in the paddies.

Floating aquaponics systems on polycultural fish ponds were installed in China in more recent years on a large scale growing rice, wheat and canna lily and other crops, with some installations exceeding 2.5 acres (10,000 m²).

The development of modern aquaponics is often attributed to the various works of the New Alchemy Institute and the works of Dr. Mark McMurtry et al. at the North Carolina State University. Inspired by the successes of the New Alchemy Institute, and the reciprocating aquaponics techniques developed by Dr. Mark McMurtry et al., other institutes soon followed suit. Starting in 1997, Dr. James Rakocy and his colleagues at the University of the Virgin Islands researched and developed the use of deep water culture hydroponic grow beds in a large-scale aquaponics system.

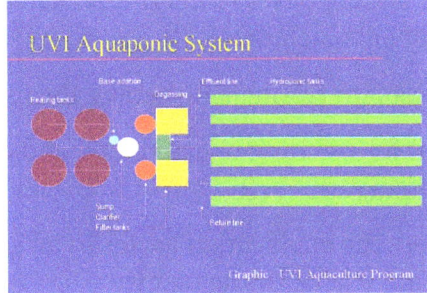

Diagram of the University of the Virgin Islands commercial aquaponics system designed to yield 5 metric tons of Tilapia per year.

The first aquaponics research in Canada was a small system added onto existing aquaculture research at a research station in Lethbridge, Alberta. Canada saw a rise in aquaponics setups throughout the '90s, predominantly as commercial installations raising high-value crops such as trout and lettuce. A setup based on the deep water system developed at the University of Virgin Islands was built in a greenhouse at Brooks, Alberta where Dr. Nick Savidov and colleagues researched aquaponics from a background of plant science. The team made findings on rapid root growth in aquaponics systems and on closing the solid-waste loop, and found that owing to certain advantages in the system over traditional aquaculture, the system can run well at a low pH level, which is favoured by plants but not fish.

Parts of an Aquaponic System

A commercial aquaponics system. An electric pump moves nutrient-rich water from the fish tank through a solids filter to remove particles the plants above cannot absorb. The water then provides nutrients for the plants and is cleansed before returning to the fish tank below.

Aquaponics consists of two main parts, with the aquaculture part for raising aquatic animals and the hydroponics part for growing plants. Aquatic effluents, resulting from uneaten feed or raising animals like fish, accumulate in water due to the closed-system recirculation of most aquaculture systems. The effluent-rich water becomes toxic to the aquatic animal in high concentrations but this contains nutrients essential for plant growth. Although consisting primarily of these two parts, aquaponics systems are usually grouped into several components or subsystems responsible for the effective removal of solid wastes, for adding bases to neutralize acids, or for maintaining water oxygenation. Typical components include:

- Rearing tank: the tanks for raising and feeding the fish;

- Settling basin: a unit for catching uneaten food and detached biofilms, and for settling out fine particulates;

- Biofilter: a place where the nitrification bacteria can grow and convert ammonia into nitrates, which are usable by the plants;

- Hydroponics subsystem: the portion of the system where plants are grown by absorbing excess nutrients from the water;

- Sump: the lowest point in the system where the water flows to and from which it is pumped back to the rearing tanks.

Depending on the sophistication and cost of the aquaponics system, the units for solids removal, biofiltration, and/or the hydroponics subsystem may be combined into one unit or subsystem, which prevents the water from flowing directly from the aquaculture part of the system to the hydroponics part.

Live Components

An aquaponic system depends on different live components to work successfully. The three main live components are plants, fish (or other aquatic creatures) and bacteria. Some systems also include additional live components like worms.

Plants

A Deep Water Culture hydroponics system where plant grow directly into the effluent rich water without a soil medium. Plants can be spaced closer together because the roots do not need to expand outwards to support the weight of the plant.

Plant placed into a nutrient rich water channel in a Nutrient film technique (NFT) system

Many plants are suitable for aquaponic systems, though which ones work for a specific system depends on the maturity and stocking density of the fish. These factors influence the concentration of nutrients from the fish effluent, and how much of those nutrients are made available to the plant roots via bacteria.

Green leaf vegetables with low to medium nutrient requirements are well adapted to aquaponic systems, including chinese cabbage, lettuce, basil, spinach, chives, herbs, and watercress.

Other plants, such as tomatoes, cucumbers, and peppers, have higher nutrient requirements and will only do well in mature aquaponic systems that have high stocking densities of fish.

Plants that are common in salads have some of the greatest success in aquaponics, including cucumbers, shallots, tomatoes, lettuce, chiles, capsicum, red salad onions and snow peas.

Some profitable plants for aquaponic systems include chinese cabbage, lettuce, basil, roses, tomatoes, okra, cantaloupe and bell peppers.

Other species of vegetables that grow well in an aquaponic system include watercress, basil, coriander, parsley, lemongrass, sage, beans, peas, kohlrabi, taro, radishes, strawberries, melons, onions, turnips, parsnips, sweet potato, cauliflower, cabbage, broccoli, and eggplant as well as the choys that are used for stir fries.

Fruiting plants like melons or tomatoes, and plants with higher nutrient needs need higher stocking densities of fish and more mature tanks to provide enough nutrients.

Fish (or Other Aquatic Creatures)

Freshwater fish are the most common aquatic animal raised using aquaponics, although freshwater crayfish and prawns are also sometimes used. There is a branch of aquaponics using saltwater fish, called saltwater aquaponics. There are many species of warmwater and coldwater fish that adapt well to aquaculture systems.

Filtered water from the hydroponics system drains into a catfish tank for re-circulation.

In practice, tilapia are the most popular fish for home and commercial projects that are intended to raise edible fish because it is a warmwater fish species that can tolerate crowding and changing water conditions. Barramundi, silver perch, eel-tailed catfish or tandanus catfish, jade perch and Murray cod are also used. For temperate climates when there isn't ability or desire to maintain water temperature, bluegill and catfish are suitable fish species for home systems.

Koi and goldfish may also be used, if the fish in the system need not be edible.

Other suitable fish include channel catfish, rainbow trout, perch, common carp, Arctic char, largemouth bass and striped bass.

Bacteria

Nitrification, the aerobic conversion of ammonia into nitrates, is one of the most important functions in an aquaponics system as it reduces the toxicity of the water for fish, and allows the resulting nitrate compounds to be removed by the plants for nourishment. Ammonia is steadily released into the water through the excreta and gills of fish as a product of their metabolism, but must be filtered out of the water since higher concentrations of ammonia (commonly between 0.5 and 1 ppm) can kill fish. Although plants can absorb ammonia from the water to some degree, nitrates are assimilated more easily, thereby efficiently reducing the toxicity of the water for fish. Ammonia can be converted into other nitrogenous compounds through combined healthy populations of:

- Nitrosomonas: bacteria that convert ammonia into nitrites, and

- Nitrobacter: bacteria that convert nitrites into nitrates.

Hydroponic Subsystem

Plants are grown as in hydroponics systems, with their roots immersed in the nutrient-rich effluent water. This enables them to filter out the ammonia that is toxic to the aquatic animals, or its metabolites. After the water has passed through the hydroponic subsystem, it is cleaned and oxygenated, and can return to the aquaculture vessels. This cycle is continuous. Common aquaponic applications of hydroponic systems include:

- Deep-water raft aquaponics: styrofoam rafts floating in a relatively deep aquaculture basin in troughs.

- Recirculating aquaponics: solid media such as gravel or clay beads, held in a container that is flooded with water from the aquaculture. This type of aquaponics is also known as closed-loop aquaponics.

- Reciprocating aquaponics: solid media in a container that is alternately flooded and drained utilizing different types of siphon drains. This type of aquaponics is also known as flood-and-drain aquaponics or ebb-and-flow aquaponics.

- Other systems use towers that are trickle-fed from the top, nutrient film technique channels, horizontal PVC pipes with holes for the pots, plastic barrels cut in half with gravel or rafts in them. Each approach has its own benefits.

Since plants at different growth stages require different amounts of minerals and nutrients, plant harvesting is staggered with seedlings growing at the same time as mature plants. This ensures stable nutrient content in the water because of continuous symbiotic cleansing of toxins from the water.

Biofilter

In an aquaponics system, the bacteria responsible for the conversion of ammonia to usable nitrates for plants form a biofilm on all solid surfaces throughout the system that are in constant contact with the water. The submerged roots of the vegetables combined have a large surface area where many bacteria can accumulate. Together with the concentrations of ammonia and nitrites in the water, the surface area determines the speed with which nitrification takes place. Care for these bacterial colonies is important as to regulate the full assimilation of ammonia and nitrite. This is why most aquaponics systems include a biofiltering unit, which helps facilitate growth of these microorganisms. Typically, after a system has stabilized ammonia levels range from 0.25 to 2.0 ppm; nitrite levels range from 0.25 to 1 ppm, and nitrate levels range from 2 to 150 ppm. During system startup, spikes may occur in the levels of ammonia (up to 6.0 ppm) and nitrite (up to 15 ppm), with nitrate levels peaking later in the startup phase. Since the nitrification process acidifies the water, non-sodium bases such as potassium hydroxide or calcium hydroxide can be added for neutralizing the water's pH if insufficient quantities are naturally present in the water to provide a buffer against acidification. In addition, selected minerals or nutrients such as iron can be added in addition to the fish waste that serves as the main source of nutrients to plants.

A good way to deal with solids buildup in aquaponics is the use of worms, which liquefy the solid organic matter so that it can be utilized by the plants and/or other animals in the system. For a worm-only growing method.

Operation

The five main inputs to the system are water, oxygen, light, feed given to the aquatic animals, and electricity to pump, filter, and oxygenate the water. Spawn or fry may be added to replace grown fish that are taken out from the system to retain a stable system. In terms of outputs, an aquaponics system may continually yield plants such as vegetables grown in hydroponics, and edible aquatic species raised in an aquaculture. Typical build ratios are .5 to 1 square foot of grow space for every 1 U.S. gal (3.8 L) of aquaculture water in the system. 1 U.S. gal (3.8 L) of water can support between .5 lb (0.23 kg) and 1 lb (0.45 kg) of fish stock depending on aeration and filtration.

Ten primary guiding principles for creating successful aquaponics systems were issued by Dr. James Rakocy, the director of the aquaponics research team at the University of the Virgin Islands, based on extensive research done as part of the Agricultural Experiment Station aquaculture program.

- Use a feeding rate ratio for design calculations

- Keep feed input relatively constant

- Supplement with calcium, potassium and iron

- Ensure good aeration

- Remove solids

- Be careful with aggregates

- Oversize pipes

- Use biological pest control

- Ensure adequate biofiltration

- Control pH

Feed Source

As in most aquaculture based systems, stock feed often consists of fish meal derived from lower-value species. Ongoing depletion of wild fish stocks makes this practice unsustainable. Organic fish feeds may prove to be a viable alternative that relieves this concern. Other alternatives include growing duckweed with an aquaponics system that feeds the same fish grown on the system, excess worms grown from vermiculture composting, using prepared kitchen scraps, as well as growing black soldier fly larvae to feed to the fish using composting grub growers.

Water Usage

Aquaponic systems do not typically discharge or exchange water under normal operation, but instead recirculate and reuse water very effectively. The system relies on the relationship between the animals and the plants to maintain a stable aquatic environment that experience a minimum of fluctuation in ambient nutrient and oxygen levels. Water is added only to replace water loss from absorption and transpiration by plants, evaporation into the air from surface water, overflow from the system from rainfall, and removal of biomass such as settled solid wastes from the system. As a result, aquaponics uses approximately 2% of the water that a conventionally irrigated farm requires for the same vegetable production. This allows for aquaponic production of both crops and fish in areas where water or fertile land is scarce. Aquaponic systems can also be used to replicate controlled wetland conditions. Constructed wetlands can be useful for biofiltration and treatment of typical household sewage. The nutrient-filled overflow water can be accumulated in catchment tanks, and reused to accelerate growth of crops planted in soil, or it may be pumped back into the aquaponic system to top up the water level.

Energy Usage

Aquaponic installations rely in varying degrees on man-made energy, technological solutions, and environmental control to achieve recirculation and water/ambient temperatures. However, if a system is designed with energy conservation in mind, using alternative energy and a reduced number of pumps by letting the water flow downwards as much as possible, it can be highly energy efficient. While careful design can minimize the risk, aquaponics systems can have multiple 'single points of failure' where problems such as an electrical failure or a pipe blockage can lead to a complete loss of fish stock.

Current Examples

Vegetable production part of the low-cost Backyard Aquaponics System developed at Bangladesh Agricultural University

The Caribbean island of Barbados created an initiative to start aquaponics systems at home, with revenue generated by selling produce to tourists in an effort to reduce growing dependence on imported food.

Dakota College at Bottineau in Bottineau, North Dakota has an aquaponics program that gives students the ability to a obtain a certificate or an AAS degree in aquaponics.

In Bangladesh, the world's most densely populated country, most farmers use agrochemicals to enhance food production and storage life, though the country lacks oversight on safe levels of chemicals in foods for human consumption. To combat this issue, a team led by M.A. Salam at the Department of Aquaculture of Bangladesh Agricultural University has created plans for a low-cost aquaponics system to provide organic produce and fish for people living in adverse climatic conditions such as the salinity-prone southern area and the flood-prone haor area in the eastern region. Salam's work innovates a form of subsistence farming for micro-production goals at the community and personal levels whereas design work by Chowdhury and Graff was aimed exclusively at the commercial level, the latter of the two approaches take advantage of economies of scale.

With more than a third of Palestinian agricultural lands in the Gaza Strip turned into a buffer zone by Israel, an aquaponic gardening system is developed appropriate for use on rooftops in Gaza City.

There has been a shift towards community integration of aquaponics, such as the nonprofit foundation Growing Power that offers Milwaukee youth job opportunities and training while growing food for their community. The model has spawned several satellite projects in other cities, such as New Orleans where the Vietnamese fisherman community has suffered from the Deepwater Horizon oil spill, and in the South Bronx in New York City.

Whispering Roots is a non-profit organization in Omaha, Nebraska that provides fresh, locally grown, healthy food for socially and economically disadvantaged communities by using aquaponics, hydroponics and urban farming.

In addition, aquaponic gardeners from all around the world are gathering in online community sites and forums to share their experiences and promote the development of this form of gardening as well as creating extensive resources on how to build home systems.

Recently, aquaponics has been moving towards indoor production systems. In cities like Chicago, entrepreneurs are utilizing vertical designs to grow food year round. These systems can be used to grow food year round with minimal to no waste.

There are various modular systems made for the public that utilize aquaponic systems to produce organic vegetables and herbs, and provide indoor decor at the same time. These systems can serve as a source of herbs and vegetables indoors. Universities are promoting research on these modular systems as they get more popular among city dwellers.

Sump

A sump (American English and some parts of Canada: oil pan) is a low space that collects often undesirable liquids such as water or chemicals. A sump can also be an infiltration basin used to manage surface runoff water and recharge underground aquifers. Sump can also refer to an area in a cave where an underground flow of water exits the cave into the earth.

Examples

One common example of a sump is the lowest point in a basement, into which flows water that seeps in from outside. If this is a regular problem, a sump pump that moves the water outside of the house may be used.

Another example is the oil pan of an engine. The oil is used to lubricate the engine's moving parts and it pools in a reservoir known as its sump, at the bottom of the engine. Use of a sump requires the engine to be mounted slightly higher to make space for it. Often though, oil in the sump can slosh during hard cornering, starving the oil pump. For these reasons, racing and piston aircraft engines are "dry sumped" using scavenge pumps and a swirl tank to separate oil from air, which is also sucked up by the pumps.

A sump can also be found in an aquarium, mainly a reef system. The sump sits below the main tank and is used as a filter, as well as a holding place of unsightly equipment such as heaters and protein skimmers. The main advantage of having a sump plumbed into an aquarium is the increase of water in the system, making it more stable and less prone to fluctuations of pH and salinity.

A diving snorkel can have a sump section located below the mouthpiece. This allows excess moisture from the breath and liquid from the ocean to settle and remain in the sump, so that it does not impair the snorkeler's breathing.

In a nuclear power plant's reactor housing, the role of the sump will be to collect any overflow of primary loop coolant; in this case, monitoring and pumping of the sump is an important part of the reactor's safety system.

The equivalent of a sump on a boat is the bilge.

In the human eye, the vitreous humour has a minor role as a metabolic sump.

Other Uses

In a foxhole, a grenade sump is a deeper hole dug inside the foxhole into which live grenades can be kicked to minimize damage from the explosion.

In medieval cosmology, the sump was the center of the cosmos, where the dregs and filth descended, with the celestial sphere far exalted above the world of fallen man.

Hydroponic Dosers

Hydroponic dosers are used by indoor, hydroponic farmers to automate the task of dispensing pH solution and/or nutrient solution as necessary.

When growing in hydroponics, the pH of the nutrient solution often drifts off target during use. The same is said for the amount of nutrients in the solution. These devices will automatically measure and adjust the solution as needed. By keeping TDS and pH levels in-range, plants grow efficiently, without health problems. Hydroponic dosers are generally specific for adjusting TDS or pH.

Ultrasonic Hydroponic Fogger

In a hydroponic system ultrasonic hydroponic foggers are used to create a fine mist, the individual particle size of which is typically of about 5 μm in diameter. These fine particles are capable of carrying nutrients from the standing water of a reservoir to plant roots. Benefits include humidification and exponentially improved root exposure to oxygen.

Ultrasonic hydroponic foggers can be used in conjunction with prior advances such as aeroponic misters or even ebb and flow systems to help improve humidity levels (mimicking a rainforest canopy) and increase nutrient absorption, thus boosting growth rates. Research into standalone fogger-supported hydroponic growing is underway.

Historical Hydroculture

This is a history of notable hydroculture phenomena. Ancient hydroculture proposed sites, and modern revolutionary works are mentioned. Included in this history are all forms of aquatic and semi-aquatic based horticulture that focus on flora: aquatic gardening, semi-aquatic crop farming, hydroponics, aquaponics, passive hydroponics, and modern aeroponics.

Hanging Gardens of Babylon

Hanging Gardens of Babylon

One of the wonders of the ancient world, was irrigated by the Euphrates River. It is uncertain if Sammu-ramat or Nebuchadrezzar II ordered them to be built between 8th and 7th century BC Babylonia. The gardens were built partially on top of ziggurats, and plants were irrigated on channels. No direct evidence of the Hanging Gardens of Babylon exists. However, there is archeological evidence, uncovered by Robert Koldewey, that ancient structures exist to support the technology used for these gardens. Ancient Greeks Diodorus Siculus and Strabo have noted the Hanging Babylonian Gardens.

Precolonial America

A Chinampa is a floating garden armada in a lake from the Xochimilco region, once Chinampan, of Mexico. This floating garden, still in use, can have an area of up to 10 meters by 200 meters.

The agricultural output of the chinampa allowed the postclassic Aztec civilization to flourish.

Historical Orient

Historically, fish have been raised within flooded rice fields in Indochina and China.

Living Root Bridges

There are 500-year-old bridges made by living roots in India, sculpted by the War-Khasis. These trees span rivers, and may be limited in connectivity to hydroculture.

Modern

Hydroculture Found in Nature

Orchids are well known for their aerial roots.

Ōhi›a Lehua, Metrosideros polymorpha, is a Hawaiian plant with roots that can grow suspended in extinct lava tubes. The roots of this plant are able to penetrate deep into the volcanic rock, to reach these hollow tubes, where they can collect moisture.

Hanging Gardens of Babylon

This hand-coloured engraving, probably made in the 19th century after the first excavations in the Assyrian capitals, depicts the fabled Hanging Gardens, with the Tower of Babel in the background

The Hanging Gardens of Babylon, one of the Seven Wonders of the Ancient World, is the only one whose location has not been definitively established.

The Hanging Gardens were described as a remarkable feat of engineering: an ascending series of tiered gardens containing all manner of trees, shrubs, and vines. The gardens were said to have looked like a large green mountain constructed of mud bricks.

Traditionally they were said to have been built in the ancient city of Babylon, near present-day Hillah, Babil province, in Iraq. The Babylonian priest Berossus, writing in about 290 BC and quoted later by Josephus, attributed the gardens to the Neo-Babylonian king Nebuchadnezzar II, who ruled between 605 and 562 BC. There are no extant Babylonian texts which mention the gardens, and no definitive archaeological evidence has been found in Babylon.

According to one legend, Nebuchadnezzar II built the Hanging Gardens for his Median wife, Queen Amytis, because she missed the green hills and valleys of her homeland. He also built a grand palace that came to be known as "The Marvel of the Mankind".

Because no physical evidence for the Hanging Gardens has be found at Babylon it has been suggested that they were purely mythical, and the descriptions found in ancient Greek and Roman writers including Strabo, Diodorus Siculus and Quintus Curtius Rufus represent a romantic ideal of an eastern garden. If it did indeed exist, it was destroyed sometime after the first century AD.

Stephanie Dalley suggests that the original garden may have been a well-documented one that the Assyrian king Sennacherib (704–681 BC) built in his capital city of Nineveh on the River Tigris, near the modern city of Mosul.

Ancient Texts

There are five principal writers whose descriptions of Babylon are extant in some form today. These writers concern themselves with the size of the Hanging Gardens, their overall design and means of irrigation, and why they were built.

Josephus (c.37–100 AD) quotes a description of the gardens by Berossus, a Babylonian priest of Marduk writing circa 290 BC. Berossus described the reign of Nebuchadnezzar II, and is the only source to credit that king with the construction of the Hanging Gardens.

In this palace he erected very high walks, supported by stone pillars; and by planting what was called a pensile paradise, and replenishing it with all sorts of trees, he rendered the prospect an exact resemblance of a mountainous country. This he did to gratify his queen, because she had been brought up in Media, and was fond of a mountainous situation.

For his description of the gardens, Diodorus Siculus (active c.60–30 BC) seems to have consulted the 4th century BC texts of both Cleitarchus (a historian of Alexander the Great) and Ctesias of Cnidus. Diodorus ascribes the construction to a Syrian king.

The park extended four plethra on each side, and since the approach to the garden sloped like a hillside and the several parts of the structure rose from one another tier on tier, the appearance of the whole resembled that of a theatre. When the ascending terraces had been built, there had been constructed beneath them galleries which carried the entire weight of the planted garden and rose little by little one above the other along the approach; and the uppermost gallery, which was fifty cubits high, bore the highest surface of the park, which was made level with the circuit wall of the battlements of the city. Furthermore, the walls, which had been constructed at great expense, were twenty-two feet thick, while the passageway between each two walls was ten feet wide. The roof above these beams had first a layer of reeds laid in great quantities of bitumen, over this two courses of baked brick bonded by cement, and as a third layer of covering of lead, to the end that the moisture from the soil might not penetrate beneath. On all this again earth had been piled to a depth sufficient for the roots of the largest trees; and the ground, when levelled off, was thickly planted with trees of every kind that, by their great size or other charm, could give pleasure to the beholder. And since the galleries, each projecting beyond another, all received the light, they contained many royal lodgings of every description; and there was one gallery which contained openings leading from the topmost surface and machines for supplying the gardens with water, the machines raising the water in great abundance from the river, although no one outside could see it being done. Now this park, as I have said, was a later construction.

The next account is from Quintus Curtius Rufus (active 1st century AD), who probably drew on the same sources as Diodorus.

The Babylonians also have a citadel twenty stades in circumference. The foundations of its turrets are sunk thirty feet into the ground and the fortifications rise eighty feet above it at the highest point. On its summit are the hanging gardens, a wonder celebrated by the fables of the Greeks. They are as high as the top of the walls and owe their charm to the shade of many tall trees. The columns supporting the whole edifice are built of rock, and on top of them is a flat surface of squared stones strong enough to bear the deep layer of earth placed upon it and the water used for irrigating it. So stout are the trees the structure supports that their trunks are eight cubits thick and their height as much as fifty feet; they bear fruit as abundantly as if they were growing in their natural environment. And although time with its gradual decaying processes is as destructive to nature's works as to man's, even so this edifice survives undamaged, despite being subjected to the pressure of so many tree-roots and the strain of bearing the weight of such a huge forest. It has a substructure of walls twenty feet thick at eleven foot intervals, so that from a distance one has the impression of woods overhanging their native mountains. Tradition has it that it is the work of a Syrian king who ruled from Babylon. He built it out of love for his wife who missed the woods and forests in this flat country and persuaded her husband to imitate nature's beauty with a structure of this kind.

Strabo (c.64 BC – 21 AD) described the Hanging Gardens as follows, in a passage thought to be based on the lost account of Onesicritus from the 4th century BC:

Babylon, too, lies in a plain; and the circuit of its wall is three hundred and eighty-five stadia. The thickness of its wall is thirty-two feet; the height thereof between the towers is fifty cubits; that of the towers is sixty cubits; and the passage on top of the wall is such that four-horse chariots can easily pass one another; and it is on this account that this and the hanging garden are called one of the Seven Wonders of the World. The garden is quadrangular in shape, and each side is four plethra in length. It consists of arched vaults, which are situated, one after another, on checkered, cube-like foundations. The checkered foundations, which are hollowed out, are covered so deep with earth that they admit of the largest of trees, having been constructed of baked brick and asphalt – the foundations themselves and the vaults and the arches. The ascent to the uppermost terrace-roofs is made by a stairway; and alongside these stairs there were screws, through which the water was continually conducted up into the garden from the Euphrates by those appointed for this purpose, for the river, a stadium in width, flows through the middle of the city; and the garden is on the bank of the river.

Philo of Byzantium (4th–5th century AD) is the last of the classical sources. The description of the gardens in his A Handbook to the Seven Wonders of the World is thought to be independent of earlier Greek sources. The method of raising water by screw matches that described by Strabo.

The so-called Hanging Gardens have plants above ground, and are cultivated in the air, with the roots of the trees above the (normal) tilled earth, forming a roof. Four stone columns are set beneath, so that the entire space through the carved pillars is beneath the (artificial) ground. Palm trees lie in place on top of the pillars, alongside each other as (cross-) beams, leaving very little space in between. This timber does not rot, unlike others; when it is soaked and put under pressure it swells up and nourishes the growth from roots, since it incorporates into its own interstices what is planted with it from outside. Much deep soil is piled on, and then broad-leaved and especially garden trees of many varieties are planted, and all kind of flowering plants, everything, in short, that is most joyous and pleasurable to the onlooker. The place is cultivated as if it were (normal) tilled earth, and the growth of new shoots has to be pruned almost as much as on normal land. This (artificial) arable land is above the heads of those who stroll along through the pillars. When the uppermost surface is walked on, the earth on the roofing stays firm and undisturbed just like a (normal) place with deep soil. Aqueducts contain water running from higher places; partly they allow the flow to run straight downhill, and partly they force it up, running backwards, by means of a screw; through mechanical pressure they force it round and round the spirals of the machines. Being discharged into close-packed, large cisterns, altogether they irrigate the whole garden, inebriating the roots of the plants to their depths, and maintaining the wet arable land, so that it is just like an evergreen meadow, and the leaves of the trees, on the tender new growth, feed upon dew and have a wind-swept appearance. For the roots, suffering no thirst, sprout anew, benefitting from the moisture of the water that runs past, flowing at random, interweaving along the lower ground to the collecting point, and reliably protects the growing of trees that have become established. Exuberant and fit for a king is the ingenuity, and most of all, forced, because the cultivator's hard work is hanging over the heads of the spectators.

Scholarship and Controversy

There is some controversy as to whether the Hanging Gardens were an actual construction or a poetic creation, owing to the lack of documentation in contemporaneous Babylonian sources. There is also no mention of Nebuchadnezzar's wife Amyitis (or any other wives), although a political marriage to a Median or Persian would not have been unusual. Many records exist of Nebuchadnezzar's works, yet his long and complete inscriptions do not mention any garden.

Herodotus, who describes Babylon in his Histories, does not mention the Hanging Gardens.

This copy of a bas relief from the North Palace of Ashurbanipal (669–631 BC) at Nineveh shows a luxurious garden watered by an aqueduct.

To date, no archaeological evidence has been found at Babylon for the Hanging Gardens. It is possible that evidence exists beneath the Euphrates, which cannot be excavated safely at present. The river flowed east of its current position during the time of Nebuchadnezzar II, and little is known about the western portion of Babylon. Rollinger has suggested that Berossus attributed the Gardens to Nebuchadnezzar for political reasons, and that he had adopted the legend from elsewhere.

A recent theory proposes that the Hanging Gardens of Babylon were actually constructed by the Assyrian king Sennacherib (reigned 704 – 681 BC) for his palace at Nineveh. Stephanie Dalley posits that during the intervening centuries the two sites became confused, and the extensive gardens at Sennacherib's palace were attributed to Nebuchadnezzar II's Babylon. Recently discovered evidence includes excavation of a vast system of aqueducts inscribed to Sennacherib, which Dalley proposes were part of a 80-kilometre (50 mi) series of canals, dams, aqueducts, used to carry water to Nineveh with water-raising screws used to raise it to the upper levels of the gardens.

Dalley bases her arguments on recent developments in the decipherment of contemporary Akkadian inscriptions. Her main points are:

- The name "Babylon", meaning "Gate of the Gods" was applied to several Mesopotamian cities. Sennacherib renamed the city gates of Nineveh after gods, which suggests that he wished his city to be considered "a Babylon".

- Only Josephus names Nebuchadnezzar as the king who built the gardens, but although Nebuchadnezzar left many inscriptions none mentions any garden or engineering works. Diodorus Siculus and Quintus Curtius Rufus specify a "Syrian" king.

- By contrast Sennacherib left written descriptions and there is archaeological evidence of his water engineering. His grandson Assurbanipal pictured the mature garden on a sculptured wall panel in his palace.

- Sennacherib called his new palace and garden "a wonder for all peoples". He describes the making and operation of screws to raise water in his garden.

- The descriptions of the classical authors fit closely to these contemporary re-cords. Before the Battle of Gaugamela in 331 BC Alexander the Great camped for four days near the aqueduct at Jerwan. The historians who travelled with him would have had ample time to investigate the enormous works around them, recording them in Greek. These first-hand accounts do not survive into our times, but were quoted by later Greek writers.

Hanging Garden at Nineveh

King Sennacherib's Hanging Garden was considered a World Wonder not just for its beauty – a year-round oasis of lush green in a dusty summer landscape – but also for the marvellous feats of water engineering that maintained the garden.

There was a tradition of Assyrian royal garden building. King Ashurnasirpal II (883–859 BC) describes what he had done:

I dug out a canal from the (river) Upper Zab, cutting through a mountain peak, and called it the Abundance Canal. I watered the meadows of the Tigris and planted orchards with all kinds of fruit trees in the vicinity. I planted seeds and plants that I had found in the countries through which I had marched and in the highlands which I had crossed: pines of different kinds, cypresses and junipers of different kinds, almonds, dates, ebony, rosewood, olive, oak, tamarisk, walnut, terebinth and ash, fir, pomegranate, pear, quince, fig, grapevine.... The canal water gushes from above into the garden; fragrance pervades the walkways, streams of water as nu-merous as the stars of heaven flow in the pleasure garden.... Like a squirrel I pick fruit in the garden of delights...

Sennacherib is the only Mesopotamian king who has left a record of his love for his wife – a key part of the romantic classical story:

And for Tashmetu-sharrat the palace woman, my beloved wife, whose features the Mis-tress of the Gods has made perfect above all other women, I had a palace of loveliness, delight and joy built...

Sennacherib's palace was comparable in size to Windsor Castle in England. He specifi-cally mentions the massive limestone blocks that reinforce the flood defences. Parts of the palace were excavated by Austin Henry Layard in the mid-19th century. His cita-del plan shows contours which would be consistent with Sennacherib's garden, but its position has not been confirmed. The area has been used as a military base in recent times, making it difficult to investigate further.

A sculptured wall panel of Assurbanipal shows the garden in its maturity. There is one original panel and the drawing of another in the British Museum, although neither is on public display. Several features mentioned by the classical authors are discernible on these contemporary images.

The irrigation of such a garden demanded an upgraded water supply to the city of Nineveh. The canals stretched over 50 km into the mountains. Sennacherib was proud of the technologies he had employed, and describes them in some detail on his inscriptions. For example:

At the headwater of Bavian (Khinnis) his inscription mentions automatic sluice gates but does not say how they worked:

The sluice gate of that canal opens without a spade or a shovel and lets the waters of abundance flow.: its sluice gate is not opened by the labour of men's hands, but by the will of the gods.

An enormous aqueduct crossing the valley at Jerwan was constructed of over 2 million dressed stones. It used stone arches and waterproof cement. On it is written:

Sennacherib king of the world king of Assyria. Over a great distance I had a watercourse directed to the environs of Nineveh, joining together the waters.... Over steep-sided valleys I spanned an aqueduct of white limestone blocks, I made those waters flow over it.

He claims to be the first to deploy a new casting technique in place of the "lost-wax" process for his monumental (30 tonne) bronze castings, and describes the making of his water screws (though once again he does not say exactly how they were driven):

Whereas in former times the kings my forefathers had created bronze statues imitating real-life forms to put on display inside temples, but in their method of work they had exhausted all the craftsmen, for lack of skill and failure to understand the principles they needed so much oil, wax and tallow for the work that they caused a shortage in their own countries – I Sennacherib, leader of all princes, knowledgeable in all kinds of work, took much advice and deep thought over doing that kind of work.... I created clay moulds as if by divine intelligence for cylinders and screws... In order to draw up water all day long, I had ropes, bronze wires and bronze chains made. And instead of a shaduf I set up the cylinders and screws of copper over cisterns....I raised the height of the surroundings of the palace, to be a Wonder for all Peoples... A high garden imitating the Amanus mountains I laid out next to it, with all kinds of aromatic plants, orchard fruit trees, trees that enrich not only mountain country but also Chaldea (Babylonia), as well as trees that bear wool, planted within it.

Sennacherib could bring the water into his garden at a high level because it was sourced from further up the mountains. He then raised the water even higher by deploying his new water screws. This meant he could build a garden that towered into the sky with large trees on the top of the terraces – a stunning artistic effect that surpassed those of his predecessors and which justifies his own claim to have built a "Wonder for all Peoples".

Chinampa

Modern chinampas

☐	Brackish Water
☐	Fresh Water
▨	Marshes
▨	*Chinampas*
—	Causeway

Mexico Valley
c. 1519

The lake system within the Valley of Mexico at the time of the Spanish Conquest, showing distribution of the chinampas.

Chinampa is a type of Mesoamerican agriculture which used small, rectangular areas of fertile arable land to grow crops on the shallow lake beds in the Valley of Mexico.

Chinampas were created by the freshwater shoreline of the Northern portion of the central lake system of Mexico by the Nahua peoples, commonly called the Aztecs. Sometimes erroneously referred to as "floating gardens", chinampas are artificial islands that were created by building up extensions of soil into bodies of water. Evidence from Nahuatl wills from late sixteenth-century Pueblo Culhuacán suggests chinampas were measured in matl (one matl=1.67 meters), often listed in groups of seven. One scholar has calculated the size of chinampas using Codex Vergara as a source, finding that they usually measured roughly 30 m × 2.5 m (98.4 ft × 8.2 ft). In Tenochtitlan, the chinampas ranged from 300 ft × 15 ft (91.4 m × 4.6 m) to 300 ft × 30 ft (91.4 m × 9.1 m) They were created by staking out the shallow lake bed and then fencing in the rectangle with wattle. The fenced-off area was then layered with mud, lake sediment, and decaying vegetation, eventually bringing it above the level of the lake. In some places, the long raised beds had ditches in between them, giving plants continuous access to water and making crops grown there independent of rainfall. Chinampas were separated by channels wide enough for a canoe to pass. These raised, well-watered beds had very high crop yields with up to 7 harvests a year. Chinampas were commonly used in pre-colonial Mexico and Central America. There is evidence that the Nahua settlement of Culhuacan, on the south side of the Ixtapalapa peninsula that divided Lake Texcoco from Lake Xochimilco, constructed the first chinampas in C.E. 1100.

History

Aztec maize agriculture as depicted in the Florentine Codex with the cultivator using a digging stick

The earliest fields that have been securely dated are from the Middle Postclassic period, 1150 – 1350 CE. Chinampas were used primarily in Lakes Xochimilco and Chalco near the springs that lined the south shore of those lakes. The Aztecs not only conducted military campaigns to obtain control over these regions but, according to some researchers, undertook significant state-led efforts to increase their extent. Chinampa farms also ringed Tenochtitlán, the Aztec capital, which was considerably enlarged over time. Smaller-scale farms have also been identified near the island-city of Xaltocan and on the east side of Lake Texcoco. With the destruction of the dams and sluice gates during the Spanish conquest of Mexico, many chinampas fields were abandoned. However, many lakeshore towns retained their chinampas through the end of the colonial era since cultivation was highly labor-intensive and less attractive for Spaniards to acquire. Xochimilco still retains a few chinampas, but in the modern era is mainly a tourist attraction.

Chinampas and canals, 1912.

Trajinera tourist boat in Xochimilco

The extent to which Tenochtitlan depended on chinampas for its fresh food supply has been the topic of a number of scholarly studies.

Among the crops grown on chinampas were maize, beans, squash, amaranth, tomatoes, chili peppers, and flowers. Maize was planted with digging stick huictli [wiktli] with a wooden blade on one end.

Chinampas were fertilized using lake sediments, likely Night soil as well, along with nutrient rich earth from the bottom of lakes. There has been scholarly work on terminology for soils and land formations.

The word chinampa comes from the Nahuatl word chināmitl, meaning "square made of canes" and the Nahuatl locative, "pan." In documentation by Spaniards, they used the word camellones, "ridges between the rows." However, Franciscan Fray Juan de Torquemada described them with the Nahua term, chinampa, saying "without much trouble [the Indians] plant and harvest their maize and greens, for all over there are ridges called chinampas; these were strips built above water and surrounded by ditches, which obviates watering."

Chinampas are depicted in pictorial Aztec codices, including Codex Vergara, Codex Santa María Asunción, the so-called Uppsala Map, the Maguey Plan (from Azcapotzalco) In alphabetic Nahuatl documentation, The Testaments of Culhuacan from the late sixteenth century have numerous references to chinampas as property that individuals bequeathed to their heirs in written wills.

There are still remnants of the chinampa system in Xochimilco, the southern portion of greater Mexico City. Chinampas have been touted as a viable model for modern for sustainable agriculture, although one anthropologist discounts that contention.

Living Root Bridges

Double living root bridge in East Khasi Hills

A living root bridge near the village of Kongthong undergoing repairs. The local War Khasis in the photo are using the young, pliable aerial roots of a fig tree to create a new railing for the bridge.

Living root bridges are a form of tree shaping common in the southern part of the Northeast Indian state of Meghalaya. They are handmade from the aerial roots of Rubber Trees (Ficus elastica) by the Khasi and Jaintia peoples of the mountainous terrain along the southern part of the Shillong Plateau.

The pliable tree roots are made to grow through betel tree trunks which have been placed across rivers and streams until the figs' roots attach themselves to the other side. Sticks, stones, and other objects are used to stabilize the growing bridge This process can take up to 15 years to complete. The useful lifespan of any given living root bridge is variable, but it is thought that, under ideal conditions, they can in principle last for many hundreds of years. As long as the tree they are formed from remains healthy, they naturally self-renew and self-strengthen as their component roots grow thicker.

Locations of Living Root Bridges

Living root bridges are known to occur in the West Jaintia Hills district and East Khasi Hills district. In the Jaintia Hills, examples of Living Root Bridges can be found in and around the villages of Shnongpdeng, Nongbareh, Khonglah, Padu, and Kudeng Rim. In the East Khasi Hills, living root bridges nearby Cherrapunji are known to exist in and around the villages of Tynrong, Mynteng, Nongriat, Nongthymmai, and around Lait-kynsew. East of Cherrapunji, examples of living root bridges are known to exist in the Khatarshnong region, in and around the villages of Nongpriang, Sohkynduh, Rymmai, Mawshuit, and Kongthong. Many more can be found near Pynursla and around the village of Mawlynnong.

History

The local Khasi people do not know when or how the tradition of living root bridges started. The earliest written record of Cherrapunji's living root bridges is by Lieutenant H Yule, who expressed astonishment about them in the 1844 Journal of the Asiatic Society of Bengal.

This living root bridge is the longest known example.

Examples of Living Root Bridges

At over 50 meters in length, the longest known example of a living root bridge is near the small Khasi town of Pynursla, India. It can be accessed from either of the villages of Mawkyrnot or Rangthylliang.

There are several examples of double living root bridges, the most famous being the "Double Decker" root bridge of Nongriat Village, pictured above.

There are three known examples of double bridges with two parallel or nearly parallel spans. Two are in the West Jaintia Hills near the villages of Padu and Nongbareh, and one is in Burma Village, in the East Khasi Hills. There is also a "Double Decker" (or possibly even "Triple Decker") near the village of Rangthylliang, close to Pynursla.

The double living root bridge of Padu Village.

Other Examples of Living Root Architecture in Meghalaya

A living root ladder near the village of Pongtung in the East Khasi Hills.

The War Khasis and War Jaintias also make several other kinds of structures out of the aerial roots of rubber trees. These include ladders and platforms. For example, in the village of Kudeng Rim in the West Jaintia Hills, a rubber tree situated next to a football field has been modified so that its branches can serve as "Living Root Bleachers." Aerial roots of the tree have been interwoven in the spaces between several branches so that platforms have been created from which villagers can watch football games.

References

- Gericke, William F. (1940). The Complete Guide to Soilless Gardening (1st ed.). London: Putnam. pp. 9–10, 38 & 84. ISBN 9781163140499.

- Douglas, James Sholto (1975). Hydroponics: The Bengal System (5th ed.). New Dehli: Oxford University Press. p. 10. ISBN 9780195605662.

- Sholto Douglas, James (1985). Advanced guide to hydroponics: (soilless cultivation). London: Pelham Books. pp. 169–187, 289–320, & 345–351. ISBN 9780720715712.

- J. Benton, Jones (2004). Hydroponics: A Practical Guide for the Soilless Grower (2nd ed.). Newyork: Taylor & Francis. pp. 29–70 & 225–229. ISBN 9780849331671.

- Dalley, Stephanie (2013). The Mystery of the Hanging Garden of Babylon: an elusive World Wonder traced. Oxford University Press. ISBN 978-0-19-966226-5.

- Dalley, Stephanie (2013) The Mystery of the Hanging Garden of Babylon: an elusive World Wonder traced, Oxford University Press ISBN 978-0-19-966226-5.

- "What Are The Easiest Plants To Grow With Aquaponics". aquaponicsideasonline.com. Retrieved 2016-01-02.

- Makoto Shinohara; Hiromi Ohmori; Yoichi Uehara. "Microbial ecosystem constructed in water for successful organic hydroponics". Retrieved 2015-01-19.

- Rogers, Patrick A. (2015-09-02). "evenfewergoats: The Undiscovered Living Root Bridges of Meghalaya Part 1: Bridges of The Umngot River Basin". evenfewergoats. Retrieved 2015-10-04.

- Rogers, Patrick A. (2015-09-14). "evenfewergoats: The Undiscovered Living Root Bridges of Meghalaya Part 2: Bridges Near Pynursla". evenfewergoats. Retrieved 2015-10-04.

- Rogers, Patrick A. (2014-01-26). "evenfewergoats: An Unknown Living Root Bridge". evenfewergoats. Retrieved 2015-10-04.

- Rogers, Patrick A. (2015-09-24). "evenfewergoats: The Undiscovered Living Root Bridges of Meghalaya Part 3: Bridges of the 12 Villages". evenfewergoats. Retrieved 2015-10-04.

- Rogers, Patrick A. (2015-09-02). "evenfewergoats: The Undiscovered Living Root Bridges of Meghalaya Part 1: Bridges of The Umngot River Basin". evenfewergoats. Retrieved 2015-10-08.

5

Environment-Friendly Gardening

Climate-friendly gardening helps in the reduction of greenhouse gasses and aids in the reduction of global warming. The alternative ways of environment friendly gardening are agroforestry, permaculture, wildlife garden, forest gardening etc. The aspects elucidated in this chapter are of vital importance and provide a better understanding of environmental friendly gardening.

Climate-Friendly Gardening

Climate-friendly gardening is gardening in ways which reduce emissions of greenhouse gases from gardens and encourage the absorption of carbon dioxide by soils and plants in order to aid the reduction of global warming. To be a climate-friendly gardener means considering both what happens in a garden and the materials brought into it and the impact they have on land use and climate. It can also include garden features or activities in the garden that help to reduce greenhouse gas emissions elsewhere.

Orchard garden showing orchard trees, herbaceous perennials and ground-cover plants, at Hergest Croft Gardens, Herefordshire, Britain.

Land use and Greenhouse Gases

Most of the excess greenhouse gases causing climate change have come from burning fossil fuel. But a special report from the Intergovernmental Panel on Climate Change (IPCC) estimated that in the last 150 years fossil fuels and cement production were responsible for only about two-thirds of climate change: the other third has been caused by human land use.

The three main greenhouse gases produced by unsustainable land use are carbon dioxide, methane, and nitrous oxide. Black carbon or soot can also be caused by unsustainable land use, and, although not a gas, can behave like greenhouse gases and contribute to climate change.

Carbon Dioxide

Carbon dioxide, CO_2, is a natural part of the carbon cycle, but human land uses often add more, especially from habitat destruction and the cultivation of soil. When woodlands, wetlands, and other natural habitats are turned into pasture, arable fields, buildings and roads, the carbon held in the soil and vegetation becomes extra carbon dioxide and methane to trap more heat in the atmosphere.

Gardeners may cause extra carbon dioxide to be added to the atmosphere in several ways:

- Using peat or potting compost containing peat;

- Buying garden furniture or other wooden products made from woodland which has been destroyed rather than taken as a renewable crop from sustainably managed woodland;

- Digging soil and leaving it bare so that the carbon in soil organic matter is oxidised;

- Using power tools which burn fossil fuel or electricity generated by burning fossil fuel;

- Using patio heaters;

- Heating greenhouses by burning fossil fuel or electricity generated by burning fossil fuel;

- Burning garden prunings and weeds on a bonfire;

- Buying tools, pesticides, synthetic nitrogen fertilizers (over 2 kilograms of carbon dioxide equivalent is produced in the manufacture of each kilogram of ammonium nitrate), and other materials which have been manufactured using fossil fuel;

- Heating and treating swimming pools by burning fossil fuel or electricity generated by burning fossil fuel;

- Watering their gardens with tapwater, which has been treated and pumped by burning fossil fuel, with a greenhouse gas impact of about 1 kg CO_2e/m^3 water.

Gardeners will also be responsible for extra carbon dioxide when they buy garden products which have been transported by vehicles powered by fossil fuel.

Methane

Methane, CH_4, is a natural part of the carbon cycle, but human land uses often add more, especially from anaerobic soil, artificial wetlands such as rice fields, and from the guts of farm animals, especially ruminants such as cattle and sheep.

Gardeners may cause extra methane to be added to the atmosphere in several ways:

- Compacting soil so that it becomes anaerobic, for example by treading on soil when it is wet;

- Allowing compost heaps to become compacted and anaerobic;

- Creating homemade liquid feed by putting the leaves of plants such as comfrey under water, with the unintended consequence that the plants may release methane as they decay;

- Killing pernicious weeds by covering them with water, with the unintended consequence that the plants may release methane as they decay;

- Allowing ponds to become anaerobic, for example by adding unsuitable fish species which stir up sediment that then blocks light from and kills submerged oxygenating plants.

Nitrous Oxide

Nitrous oxide, N_2O, is a natural part of the nitrogen cycle, but human land uses often add more.

Gardeners may cause extra nitrous oxide to be added to the atmosphere by:

- Using synthetic nitrogen fertilizer, for example "weed and feed" on lawns, especially if it is applied when plants are not actively growing, the soil is compacted, or when other factors are limiting so that the plants cannot make use of the nitrogen;

- Compacting the soil (for example by working in the garden when the soil is wet) which will increase the conversion of nitrates to nitrous oxide by soil bacteria;

- Burning garden waste on bonfires.

Black Carbon

Black carbon is not a gas, but it acts like a greenhouse gas because it can be suspended in the atmosphere and absorb heat.

Gardeners may cause extra black carbon to be added to the atmosphere by burning garden prunings and weeds on bonfires, especially if the waste is wet and becomes black carbon in the form of soot. Gardeners will also be responsible for extra black carbon produced when they buy garden products which have been transported by vehicles powered by fossil fuel especially the diesel used in most lorries.

Gardening to Reduce Greenhouse Gas Emissions and Absorb Carbon Dioxide

There are many ways in which climate-friendly gardeners may reduce their contribution to climate change and help their gardens absorb carbon dioxide from the atmosphere.

Climate-friendly gardeners can find good ideas in many other sustainable approaches:

- Agroforestry;
- Forest gardening;
- Orchards;
- Organic gardening;
- Permaculture;
- Rain garden;
- Vegan organic gardening;
- Water-wise gardening;
- Wildlife garden.

Protecting and Enhancing Carbon Stores

Protecting Carbon Stores in Land Beyond Gardens

Woodland and wetland in the New Forest, Hampshire

Woodland and trees in Herefordshire

Kitchen garden at Charles Darwin's home, Down House, Kent, showing greenhouse, waterbutt, box hedging and vegetable beds.

Alliums, lavender, box and other water-thrifty plants in the dry garden at Cambridge Botanic Garden

Climate-friendly gardening includes actions which protect carbon stores beyond gardens. The biggest carbon stores in land are in soil; the two habitat types with the biggest carbon stores per hectare are woods and wetlands; and woods absorb more carbon dioxide per hectare per year than most other habitats. Climate-friendly gardeners therefore aim to ensure that nothing they do will harm these habitats.

According to Morison and Morecroft (eds.)'s Plant Growth and Climate Change, the net primary productivity (the net amount of carbon absorbed each year) of various habitats is:

- Tropical forests: 12.5 tonnes of carbon per hectare per year;

- Temperate forests: 7.7 tonnes of carbon per hectare per year;

- Temperate grasslands: 3.7 tonnes of carbon per hectare per year;

- Croplands: 3.1 tonnes of carbon per hectare per year.

The Intergovernmental Panel on Climate Change's Special Report Land Use, Land-Use Change and Forestry lists the carbon contained in different global habitats as:

- Wetlands: 643 tonnes carbon per hectare in soil + 43 tonnes carbon per hectare in vegetation = total 686 tonnes carbon per hectare;

- Tropical forests: 123 tonnes carbon per hectare in soil + 120 tonnes carbon per hectare in vegetation = total 243 tonnes carbon per hectare;

- Temperate forests: 96 tonnes carbon per hectare in soil + 57 tonnes carbon per hectare in vegetation = total 153 tonnes carbon per hectare;

- Temperate grasslands: 164 tonnes carbon per hectare in soil + 7 tonnes carbon per hectare in vegetation = total 171 tonnes carbon per hectare;

- Croplands: 80 tonnes carbon per hectare in soil + 2 tonnes carbon per hectare in vegetation = total 82 tonnes carbon per hectare.

The figures quoted above are global averages. More recent research in 2009 has found that the habitat with the world's highest known total carbon density - 1,867 tonnes of carbon per hectare - is temperate moist forest of Eucalyptus regnans in the Central Highlands of south-east Australia; and, in general, that temperate forests contain more carbon than either boreal forests or tropical forests.

Carbon Stores in Britain

According to Milne and Brown's 1997 paper "Carbon in the vegetation and soils of Great Britain", Britain's vegetation and soil are estimated to contain 9952 million tonnes of carbon, of which almost all is in the soil, and most in Scottish peatland soil:

- Soils in Scotland: 6948 million tonnes carbon;

- Soils in England and Wales: 2890 million tonnes carbon;

- Vegetation in British woods and plantations (which cover only 11% of Britain's land area): 91 million tonnes carbon;

- Other vegetation: 23 million tonnes carbon.

A 2005 report suggested that British woodland soil may contain as much as 250 tonnes of carbon per hectare.

Many studies of soil carbon only study the carbon in the top 30 centimetres, but soil is often much deeper than that, especially below woodland. One 2009 study of the United

Kingdom's carbon stores by Keith Dyson and others gives figures for soil carbon down to 100 cm below the habitats, including "Forestland", "Cropland" and "Grassland", covered by the Kyoto Protocol reporting requirements.

- Forestland soils: average figures in tonnes carbon per hectare are 160 (England), 428 (Scotland), 203 (Wales), and 366 (Northern Ireland).

- Grassland soils: average figures in tonnes carbon per hectare are 148 (England), 386 (Scotland), 171 (Wales), and 304 (Northern Ireland).

- Cropland soils: average figures in tonnes carbon per hectare are 110 (England), 159 (Scotland), 108 (Wales), and 222 (Northern Ireland).

Protecting Carbon Stores in Wetland

Permeable paving of wood chip with birch-log edging at the Royal Horticultural Society garden at Wisley

A ground-cover and rain-garden plant - Symphytum grandiflorum, creeping comfrey (with Cotinus coggygria)

Climate-friendly gardeners choose peat-free composts because some of the planet's biggest carbon stores are in soil, and especially in the peatland soil of wetlands.

The Intergovernmental Panel on Climate Change's Special Report Land Use, Land-Use Change and Forestry gives a figure of 2011 gigatonnes of carbon for global carbon stocks in the top 1 metre of soils, much more than the carbon stores in the vegetation or the atmosphere.

Climate-friendly gardeners also avoid using tapwater not only because of the greenhouse gases emitted when fossil fuels are burnt to treat and pump water, but because if water is taken from wetlands then carbon stores are more likely to be oxidised to carbon dioxide.

A climate-friendly garden therefore does not contain large irrigated lawns, but instead includes water-butts to collect rainwater; water-thrifty plants which survive on rainwater and do not need watering after they are established; trees, shrubs and hedges to shelter gardens from the drying effects of sun and wind; and groundcover plants and organic mulch to protect the soil and keep it moist.[p. 242p. 80–82]

Climate-friendly gardeners will ensure that any paved surfaces in their gardens (which are kept to a minimum to increase carbon stores) are permeable, and may also make rain gardens, sunken areas into which rainwater from buildings and paving is directed, so that the rain can then be fed back into groundwater rather than going into storm drains. The plants in rain gardens must be able to grow in both dry and wet soils.

Protecting Carbon Stores in Woodland

Wetlands may store the most carbon in their soils, but woods store more carbon in their living biomass than any other type of vegetation, and their soils store the most carbon after wetlands. Climate-friendly gardeners therefore ensure that any wooden products they buy, such as garden furniture, have been made of wood from sustainably managed woodland.

Protecting and Increasing Carbon Stores in Gardens

Juglans elaeopyren, an American walnut, at Cambridge Botanic Garden

After rocks containing carbonate compounds, soil is the biggest store of carbon on land. Carbon is found in soil organic matter, including living organisms (plant roots, fungi, animals, protists, bacteria), dead organisms, and humus. One study of the environmental benefits of gardens estimates that 86% of carbon stores in gardens is in the soil.

Wild strawberries in flower below a British hedge.

The first priorities for climate-friendly gardeners are, therefore, to:

- Protect the soil's existing carbon stores;

- Increase the soil's carbon stores.

To protect the soil, climate-friendly gardens:

- Are based on plants rather than buildings and paving;

- Have soil that is kept at a relatively stable temperature by shelter from trees, shrubs and/or hedges;

- Have soil that is always kept covered and therefore moist and at a relatively stable temperature by groundcover plants, fast-growing green manures (which can be used as an intercrop in kitchen gardens of annual vegetables) and/or organic mulches.

Mulch of woodchips protecting soil at the Royal Horticultural Society garden at Wisley in Surrey.

Climate-friendly gardeners avoid things which may harm soil. They do not tread on the soil when it is wet, because it is then most vulnerable to compaction. They dig as little is possible, and only when the soil is moist rather than wet, because cultivation increases the oxidation of soil organic matter and produces carbon dioxide.

To increase soil carbon stores, climate-friendly gardeners ensure that their gardens create optimal conditions for vigorous healthy growth of plants, and other garden organisms above and below ground, and reduce the impact of any limiting factors.

In general, the more biomass that the plants can create each year, the more carbon will be added to the soil. However, only some biomass each year becomes long-term soil carbon or humus. In Soil Carbon and Organic Farming, a 2009 report for the Soil Association, Gundula Azeez discusses several factors which increase how much biomass is turned into humus. These include good soil structure, soil organisms such as fine root hairs, microorganisms, mycorrhizas and earthworms which increase soil aggregation, residues from plants (such as trees and shrubs) which have a high content of resistant chemicals such as lignin, and plant residues with a carbon to nitrogen ratio lower than about 32:1.

Nitrogen-fixing nodules on Wisteria roots (hazelnut for scale)

Climate-friendly gardens therefore include:

- Hedges for shelter from wind;

- A light canopy of late-leafing deciduous trees to let in enough sunlight for growth but not so much that the garden becomes too hot and dry (this is one of the principles behind many agroforestry systems, such as Paulownia's use in China partly because it is late-leafing and its canopy is sparse so that crops below it get shelter but also enough light);

- Groundcover plants and organic mulches (such as woodchips over compost made from kitchen and garden "waste") to keep soil moist and at relatively stable temperatures;

- Nitrogen-fixing plants, because soil nitrogen may be a limiting factor (but climate-friendly gardeners avoid synthetic nitrogen fertilizers, because these may cause mycorrhizal associations to break down);

- Many layers of plants, including woody plants such as trees and shrubs, other perennials, groundcover plants, deep-rooted plants, all chosen according to 'right plant, right place', so that they are suited to their growing conditions and will grow well;

- A wide diversity of disease-resistant, vigorous plants for resilience and to make the most of all available ecological niches;

- Plants to feed and shelter wildlife, to increase total biomass, and to ensure biological control of pests and diseases.

- Compost made from garden and kitchen "waste".

Lawns, like other grasslands, can build up good levels of soil carbon, but they will grow more vigorously and store more carbon if besides grasses they also contain nitrogen-fixing plants such as clover, and if they are cut using a mulching mower which returns finely-chopped mowings to the lawn. More carbon, however, may be stored by other perennial plants such as trees and shrubs. They also do not need to be maintained using power tools.

Climate-friendly gardeners will also aim to increase biodiversity not only for the sake of the wildlife itself, but so that the garden ecosystem is resilient and more likely to store as much carbon as possible as long as possible. They will therefore avoid pesticides, and increase the diversity of the habitats within their gardens.

Reducing Greenhouse Gas Emissions

Climate-friendly gardeners can directly reduce the greenhouse gas emissions from their own gardens, but can also use their gardens to indirectly reduce greenhouse gas emissions elsewhere.

Using Gardens to Reduce Greenhouse Gas Emissions

Climate-friendly gardeners can use their gardens in ways which reduce greenhouse gases elsewhere, for example by using the sun and wind to dry washing on washing lines in the garden instead of using electricity generated by fossil fuel to dry washing in tumble dryers.

From Farmland

Walnut, Juglans regia, with ripening walnuts

Food is a major contributor to climate change. In the United Kingdom, according to Tara Garnett of the Food Food Climate Research Network, food contributes 19% of the country's greenhouse gas emissions.

Soil is the biggest store of carbon on land. It is therefore important to protect the soil organic matter in farmland. Farm animals, however, especially free-range pigs, may cause erosion, and cultivation of the soil increases the oxidation of soil organic matter into carbon dioxide. Other sources of greenhouse gases from farmland include: compaction caused by farm machinery or overgrazing by farm animals can make soil anaerobic and produce methane; farm animals produce methane; and nitrogen fertilizers can be converted to nitrous oxide.

Most farmland consists of fields growing annual arable crops which are eaten directly by people or fed to farm animals, and grassland used as pasture, hay or silage to feed farm animals. Some perennial food plants are also grown, such as fruits and nuts in orchards, and watercress grown in water.

Although all cultivation of the soil in arable fields produces carbon dioxide, some arable crops cause more damage to soil than others. Root crops such as potatoes and sugar-beet, and crops which are harvested not just once a year but over a long period such as green vegetables and salads, are considered "high risk" in catchment-sensitive farming.

Climate-friendly gardeners therefore grow at least some of their food, and may choose food crops which therefore help to keep carbon in farmland soils if they grow such high-risk crops in small vegetable plots in their gardens, where it is easier to protect the soil than in large fields under commercial pressures. Climate-friendly gardeners may grow and eat plants such as sweet cicely which sweeten food, and so reduce the land area needed for sugar-beet. They may also choose to grow perennial food plants to not only reduce their indirect greenhouse gas emissions from farmland, but also to increase carbon stores in their own gardens.

Grassland contains more carbon per hectare than arable fields, but farm animals, especially ruminants such as cattle or sheep, produce large amounts of methane, directly and from manure heaps and slurry. Slurry and manure may also produce nitrous oxide. Gardeners who want to reduce their greenhouse gas emissions can help themselves to eat less meat and dairy produce by growing nut trees which are a good source of tasty, protein-rich food, including walnuts which are an excellent source of the omega-3 fatty acid alpha-linolenic acid.

Researchers and farmers are investigating and improving ways of farming which are more sustainable, such as agroforestry, forest farming, wildlife-friendly farming, soil management, catchment-sensitive farming (or water-friendly farming). For example, the organisation Farming Futures assists farmers in the United Kingdom to reduce their farms' greenhouse gas emissions.

Farmers are aware that consumers are increasingly asking for "green credentials". Gardeners who understand climate-friendly practices can advocate their use by farmers.

From Industry

Climate-friendly gardeners aim to reduce their consumption in general. In particular, they try to avoid or reduce their consumption of tapwater because of the greenhouse gases emitted when fossil fuels are burnt to supply the energy needed to treat and pump it to them. Instead, gardeners can garden using only rainwater.

Nitrogen-fixing and edible - Elaeagnus umbellatus at the Agroforestry Research Trust forest garden in Devon

Greenhouse gases are produced in the manufacture of many materials and products used by gardeners. For example, it takes a lot of energy to produce synthetic fertilizers, especially nitrogen fertilizers. Ammonium nitrate, for example, has an embodied energy of 67000 kilojoules/kilogramme, so climate-friendly gardeners will choose alternative ways of ensuring the soil in their gardens has optimal levels of nitrogen by alternative means such as nitrogen-fixing plants.

Climate-friendly gardeners will also aim to follow "cradle-to-cradle design" and "circular economy" principles: when they choose to buy or make something, it should be possible to take it apart again and recycle or compost every part, so that there is no waste, only raw materials to be made into something else. This will reduce the greenhouse gases otherwise produced when extracting raw materials.

From Transport

Gardeners can reduce not only their food miles by growing some of their own food, but also their "gardening miles" by reducing the amount of plants and other materials they import, obtaining them as locally as possible and with as little packaging as possible. This might include ordering plants by mail order from a specialist nursery if the plants are sent out bare-root, reducing transport demand and the use of peat-based composts; or growing plants from seed, which will also increase genetic diversity and therefore resilience; or growing plants vegetatively from cuttings or offsets from other local gardeners; or buying reclaimed materials from salvage firms.

From Houses

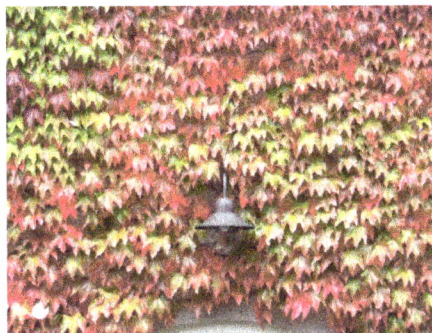

Climbers as insulation - Boston ivy, Parthenocissus tricuspidata, Boston ivy, in autumn

Climate-friendly gardeners can use their gardens in ways which reduce greenhouse gas emissions from homes by:

- Using sunlight and wind to dry washing on washing lines instead of fossil fuel-generated electricity to run tumble dryers;

- Planting deciduous climbers on houses and planting deciduous trees at suitable distances from the house to provide shade during the summer, reducing the consumption of electricity for air conditioning, but also such that at cooler times of year, sunlight can reach and warm a house, reducing heating costs and consumption;

- Planting hedges, trees, shrubs and climbers to shelter houses from wind, reducing heating costs and consumption during the winter (as long as any planting does not create a wind-tunnel effect).

Climate-friendly gardeners may also choose to reduce their own personal greenhouse gas emissions by growing and eating carminative plants such as fennel and garlic which reduce intestinal gases such as methane.

Reducing Greenhouse Gas Emissions from Gardens

Slow-growing yew, Taxus baccata, as hedge at Charles Darwin's home, Down House, Kent

Nitrogen-fixing red and white clover (Trifolium) as lawn plants

Leaf cage, compost heap and wormery at the Royal Horticultural Society garden at Wisley

There are some patent sources of greenhouse gas emissions in gardens and some more latent.

Power tools which are powered by diesel or petrol, or electricity generated by burning other fossil fuels, emit carbon dioxide. Climate-friendly gardeners may therefore choose to use hand tools rather than power tools, or power tools powered by renewable electricity, or design their gardens to reduce or remove a need to use power tools. For example, they may choose dense, slow-growing species for hedges so that the hedges only need to be cut once a year.

Lawns need[?] to be cut by lawn mowers and, in drier parts of the world, are often irrigated by tapwater. Climate-friendly gardeners will therefore do what they can to reduce this consumption by:

- Replacing part of or all lawns with other perennial planting such as trees and shrubs with less ecologically demanding maintenance requirements;

- Cut some or all lawns only once or twice a year, i.e. convert them into meadows;

- Make lawn shapes simple so that they may be cut quickly;

- Increase the cutting height of mower blades;

- Use a mulching mower to return organic matter to the soil;

- Sow clover to increase vigour (without the need for synthetic fertilisers) and resilience in dry periods;

- Cut lawns with electric mowers using electricity from renewable energy;

- Cut lawns with hand tools such as push mowers or scythes.

Greenhouses can be used to grow crops which might otherwise be imported from warmer climates, but if they are heated by fossil fuel then they may cause more greenhouse gas emissions than they save. Climate-friendly gardeners will therefore use their greenhouses carefully by:

- Choosing only annual plants which will only be in the greenhouse during warmer months, or perennial plants which do not need any extra heat during winter;

- Using water tanks as heat stores and compost heaps as heat sources inside greenhouses so that they stay frost-free in winter.

Climate-friendly gardeners will not put woody prunings on bonfires, which will emit carbon dioxide and black carbon, but instead burn them indoors in a wood-burning stove and therefore cut emissions from fossil fuel, or cut them up to use as mulch and increase soil carbon stores, or add the smaller prunings to compost heaps to keep them aerated, reducing methane emissions. To reduce the risk of fire, they will also choose fire-resistant plants from habitats which are not prone to wildfires and which do not catch fire easily, rather than fire-adapted plants from fire-prone habitats which are flammable and adapted to encourage fires and then gain a competitive advantage over less resistant species.

Climate-friendly gardeners may use deep-rooted plants such as comfrey to bring nutrients closer to the surface topsoil, but will do so without making the leaves into a liquid feed, because the rotting leaves in the anaerobic conditions under water may emit methane.

Nitrogen fertilizers may be oxidised to nitrous oxide, especially if fertilizer is applied in excess, or when plants are not actively growing. Climate-friendly gardeners may choose instead to use nitrogen-fixing plants which will add nitrogen to the soil without increasing nitrous oxide emissions.

Agroforestry

Agroforestry or agro-sylviculture is a land use management system in which trees or shrubs are grown around or among crops or pastureland. It combines shrubs and trees in agricultural and forestry technologies to create more diverse, productive, profitable, healthy, ecologically sound, and sustainable land-use systems.

Parkland in Burkina Faso: sorghum grown under Faidherbia albida and Borassus akeassii near Banfora

As a Science

The theoretical base for agroforestry comes from ecology, via agroecology. From this perspective, agroforestry is one of the three principal land-use sciences. The other two are agriculture and forestry.

Agroforestry has a lot in common with intercropping. Both have two or more plant species (such as nitrogen-fixing plants) in close interaction, both provide multiple outputs, as a consequence, higher overall yields and, because a single application or input is shared, costs are reduced. Beyond these, there are gains specific to agroforestry.

Benefits

Agroforestry systems can be advantageous over conventional agricultural, and forest production methods. They can offer increased productivity, economic benefits, and more diversity in the ecological goods and services provided .(An example of this was seen in trying to conserve Milicia excelsa.)

Biodiversity in agroforestry systems is typically higher than in conventional agricultural systems. With two or more interacting plant species in a given land area, it creates a more complex habitat that can support a wider variety of birds, insects, and other animals. Depending upon the application, impacts of agroforestry can include:

- Reducing poverty through increased production of wood and other tree products for home consumption and sale

- Contributing to food security by restoring the soil fertility for food crops

- Cleaner water through reduced nutrient and soil runoff

- Countering global warming and the risk of hunger by increasing the number of drought-resistant trees and the subsequent production of fruits, nuts and edible oils

- Reducing deforestation and pressure on woodlands by providing farm-grown fuelwood

- Reducing or eliminating the need for toxic chemicals (insecticides, herbicides, etc.)

- Through more diverse farm outputs, improved human nutrition

- In situations where people have limited access to mainstream medicines, providing growing space for medicinal plants

- Increased crop stability

- Multifunctional site use i.e. crop production and animal grazing.

- Typically more drought resistant.

- Stabilises depleted soils from erosion

- Bioremediation

Agroforestry practices may also realize a number of other associated environmental goals, such as:

- Carbon sequestration

- Odour, dust, and noise reduction

- Green space and visual aesthetics

- Enhancement or maintenance of wildlife habitat

Adaptation to Climate Change

There is some evidence that, especially in recent years, poor smallholder farmers are turning to agroforestry as a mean to adapt to the impacts of climate change. A study from the CGIAR research program on Climate Change, Agriculture and Food Security (CCAFS) found from a survey of over 700 households in East Africa that at least 50% of those households had begun planting trees on their farms in a change from their practices 10 years ago. The trees ameliorate the effects of climate change by helping to stabilize erosion, improving water and soil quality and providing yields of fruit, tea, coffee, oil, fodder and medicinal products in addition to their usual harvest. Agroforestry was one of the most widely adopted adaptation strategies in the study, along with the use of improved crop varieties and intercropping.

Applications

Agroforestry represents a wide diversity in application and in practice. One listing includes over 50 distinct uses. The 50 or so applications can be roughly classified under a

few broad headings. There are visual similarities between practices in different categories. This is expected as categorization is based around the problems addressed (countering winds, high rainfall, harmful insects, etc.) and the overall economic constraints
and objectives (labor and other inputs costs, yield requirements, etc.). The categories
include :

- Parklands
- Shade systems
- Crop-over-tree systems
- Alley cropping
- Strip cropping
- Fauna-based systems
- Boundary systems
- Taungyas
- Physical support systems
- Agroforests
- Wind break and shelterbelt.

Parkland

Parklands are visually defined by the presence of trees widely scattered over a large
agricultural plot or pasture. The trees are usually of a single species with clear regional favorites. Among the beaks and benefits, the trees offer shade to grazing animals,
protect crops against strong wind bursts, provide tree prunings for firewood, and are a
roost for insect or rodent-eating birds.

There are other gains. Research with Faidherbia albida in Zambia showed that mature trees can sustain maize yields of 4.1 tonnes per hectare compared to 1.3 tonnes
per hectare without these trees. Unlike other trees, Faidherbia sheds its nitrogen-rich
leaves during the rainy crop growing season so it does not compete with the crop for
light, nutrients and water. The leaves then regrow during the dry season and provide
land cover and shade for crops.

Shade Systems

With shade applications, crops are purposely raised under tree canopies and within the
resulting shady environment. For most uses, the understory crops are shade tolerant or
the overstory trees have fairly open canopies. A conspicuous example is shade-grown
coffee. This practice reduces weeding costs and improves the quality and taste of the

coffee. Just because plants are grown under shade does not necessarily translate into lost or reduced yields. This is because the efficiency of photosynthesis drops off with increasing light intensity, and the rate of photosynthesis hardly increases once the light intensity is over about one tenth that of direct overhead sun. This means that plants under trees can still grow well even though they get less light. By having more than one level of vegetation, it is possible to get more photosynthesis, and overall yields, than with a single canopy layer.

Crop-over-tree Systems

Not commonly encountered, crop-over-tree systems employ woody perennials in the role of a cover crop. For this, small shrubs or trees pruned to near ground level are utilized. The purpose, as with any cover crop, is to increase in-soil nutrients and/or to reduce soil erosion.

Alley Cropping

Alley cropping corn fields between rows of walnut trees.

With alley cropping, crop strips alternate with rows of closely spaced tree or hedge species. Normally, the trees are pruned before planting the crop. The cut leafy material is spread over the crop area to provide nutrients for the crop. In addition to nutrients, the hedges serve as windbreaks and eliminate soil erosion.

Alley cropping has been shown to be advantageous in Africa, particularly in relation to improving maize yields in the sub-Saharan region. Use here relies upon the nitrogen fixing tree species Sesbania sesban, euphorbia tricalii, Tephrosia vogelii, Gliricidia sepium and Faidherbia albida. In one example, a ten-year experiment in Malawi showed that, by using the fertilizer tree Gliricidia (Gliricidia sepium) on land on which no mineral fertilizer was applied, maize yields averaged 3.3 tonnes per hectare as compared to one tonne per hectare in plots without fertilizer trees nor mineral fertilizers.

Strip Cropping

Strip cropping is similar to alley cropping in that trees alternate with crops. The difference is that, with alley cropping, the trees are in single row. With strip cropping, the trees or shrubs are planted in wide strip. The purpose can be, as with alley cropping, to provide nutrients, in leaf form, to the crop. With strip cropping, the trees can have a purely productive role, providing fruits, nuts, etc. while, at the same time, protecting nearby crops from soil erosion and harmful winds.

Fauna-based Systems

~ 1970   2004

Silvopasture over the years (Australia).

There are situations where trees benefit fauna. The most common examples are the silvopasture where cattle, goats, or sheep browse on grasses grown under trees. In hot climates, the animals are less stressed and put on weight faster when grazing in a cooler, shaded environment. Other variations have these animals directly eating the leaves of trees or shrubs.

There are similar systems for other types of fauna. Deer and hogs gain when living and feeding in a forest ecosystem, especially when the tree forage suits their dietary needs. Another variation, aquaforestry, is where trees shade fish ponds. In many cases, the fish eat the leaves or fruit from the trees.

Boundary Systems

A riparian buffer bordering a river in Iowa.

There are a number of applications that fall under the heading of a boundary system. These include the living fences, the riparian buffer, and windbreaks.

- A living fence can be a thick hedge or fencing wire strung on living trees. In addition to restricting the movement of people and animals, living fences offer habitat to insect-eating birds and, in the case of a boundary hedge, slow soil erosion.

- Riparian buffers are strips of permanent vegetation located along or near active watercourses or in ditches where water runoff concentrates. The purpose is to keep nutrients and soil from contaminating surface water.

- Windbreaks reduce the velocity of the winds over and around crops. This increases yields through reduced drying of the crop and/or by preventing the crop from toppling in strong wind gusts.

Taungya

Taungya is a vastly used system originating in Burma. In the initial stages of an orchard or tree plantation, the trees are small and widely spaced. The free space between the newly planted trees can accommodate a seasonal crop. Instead of costly weeding, the underutilized area provides an additional output and income. More complex taungyas use the between-tree space for a series of crops. The crops become more shade resistant as the tree canopies grow and the amount of sunlight reaching the ground declines. If a plantation is thinned in the latter stages, this opens further the between-tree cropping opportunities.

Physical Support Systems

In the long history of agriculture, trellises are comparatively recent. Before this, grapes and other vine crops were raised atop pruned trees. Variations of the physical support theme depend upon the type of vine. The advantages come through greater in-field biodiversity. In many cases, the control of weeds, diseases, and insect pests are primary motives.

Agroforests

These are widely found in the humid tropics and are referenced by different names (forest gardening, forest farming, tropical home gardens and, where short-statured trees or shrubs dominate, shrub gardens). Through a complex, diverse mix of trees, shrubs, vines, and seasonal crops, these systems achieve the ecological dynamics of a forest ecosystem. Because of their internal ecology, they tend to be less susceptible to harmful insects, plant diseases, drought, and wind damage.

Historical Use

Agroforestry similar methods were historically utilized by Native Americans. California Indians would prescribe burn oak and other habitats to maintain a 'pyrodiversity collecting model'. This method allowed for greater health of trees and the habitat in general.

Challenges

Agroforestry is relevant to almost all environments and is a potential response to common problems around the globe, and agroforestry systems can be advantageous compared to conventional agriculture or forestry. Yet agroforestry is not very widespread, at least according to current but incomplete USDA surveys as of November, 2013.

As suggested by a survey of extension programs in the United States, some obstacles (ordered most critical to least critical) to agroforestry adoption include:

- Lack of developed markets for products
- Unfamiliarity with technologies
- Lack of awareness of successful agroforestry examples
- Competition between trees, crops, and animals
- Lack of financial assistance
- Lack of apparent profit potential
- Lack of demonstration sites
- Expense of additional management
- Lack of training or expertise
- Lack of knowledge about where to market products
- Lack of technical assistance
- Cannot afford adoption or start up costs, including costs of time
- Unfamiliarity with alternative marketing approaches (e.g. web)
- Unavailability of information about agroforestry
- Apparent inconvenience
- Lack of infrastructure (e.g. buildings, equipment)
- Lack of equipment
- Insufficient land
- Lack of seed/seedling sources
- Lack of scientific research

Some solutions to these obstacles have already been suggested although many depend on particular circumstances which vary from one location to the next.

Permaculture

Permaculture is a system of agricultural and social design principles centered on simulating or directly utilizing the patterns and features observed in natural ecosystems. Permaculture was developed, and the term coined by Bill Mollison and David Holmgren in 1978.

It has many branches that include but are not limited to ecological design, ecological engineering, environmental design, construction and integrated water resources management that develops sustainable architecture, and regenerative and self-maintained habitat and agricultural systems modeled from natural ecosystems.

Mollison has said: "Permaculture is a philosophy of working with, rather than against nature; of protracted and thoughtful observation rather than protracted and thoughtless labor; and of looking at plants and animals in all their functions, rather than treating any area as a single product system."

History

In 1929, Joseph Russell Smith took up an antecedent term as the subtitle for Tree Crops: A Permanent Agriculture, a book in which he summed up his long experience experimenting with fruits and nuts as crops for human food and animal feed. Smith saw the world as an inter-related whole and suggested mixed systems of trees and crops underneath. This book inspired many individuals intent on making agriculture more sustainable, such as Toyohiko Kagawa who pioneered forest farming in Japan in the 1930s.

The definition of permanent agriculture as that which can be sustained indefinitely was supported by Australian P. A. Yeomans in his 1964 book Water for Every Farm. Yeomans introduced an observation-based approach to land use in Australia in the 1940s, and the keyline design as a way of managing the supply and distribution of water in the 1950s.

Stewart Brand's works were an early influence noted by Holmgren. Other early influences include Ruth Stout and Esther Deans, who pioneered no-dig gardening, and Masanobu Fukuoka who, in the late 1930s in Japan, began advocating no-till orchards, gardens, and natural farming.

In the late 1960s, Bill Mollison and David Holmgren started developing ideas about stable agricultural systems on the southern Australian island state of Tasmania. This

was a result of the danger of the rapidly growing use of industrial-agricultural methods. In their view, these methods were highly dependent on non-renewable resources,
and were additionally poisoning land and water, reducing biodiversity, and removing
billions of tons of topsoil from previously fertile landscapes. A design approach called
permaculture was their response and was first made public with the publication of their
book Permaculture One in 1978.

Bill Mollison in January 2008.

By the early 1980s, the concept had broadened from agricultural systems design towards sustainable human habitats. After Permaculture One, Mollison further refined
and developed the ideas by designing hundreds of permaculture sites and writing more
detailed books, notably Permaculture: A Designers Manual. Mollison lectured in over
80 countries and taught his two-week Permaculture Design Course (PDC) to many
hundreds of students. Mollison "encouraged graduates to become teachers themselves
and set up their own institutes and demonstration sites. This multiplier effect was critical to permaculture's rapid expansion."

Core Tenets and Principles of Design

The three core tenets of permaculture are:

- Care for the earth: Provision for all life systems to continue and multiply. This
  is the first principle, because without a healthy earth, humans cannot flourish.

- Care for the people: Provision for people to access those resources necessary for
  their existence.

- Return of surplus: Reinvesting surpluses back into the system to provide for the
  first two ethics. This includes returning waste back into the system to recycle
  into usefulness. The third ethic is sometimes referred to as Fair Share to reflect
  that each of us should take no more than what we need before we reinvest the
  surplus.

Permaculture design emphasizes patterns of landscape, function, and species assem-
blies. It determines where these elements should be placed so they can provide maximum benefit to the local environment. The central concept of permaculture is maxi-

mizing useful connections between components and synergy of the final design. The focus of permaculture, therefore, is not on each separate element, but rather on the relationships created among elements by the way they are placed together; the whole becoming greater than the sum of its parts. Permaculture design therefore seeks to minimize waste, human labor, and energy input by building systems with maximal benefits between design elements to achieve a high level of synergy. Permaculture designs evolve over time by taking into account these relationships and elements and can become extremely complex systems that produce a high density of food and materials with minimal input.

The design principles which are the conceptual foundation of permaculture were derived from the science of systems ecology and study of pre-industrial examples of sustainable land use. Permaculture draws from several disciplines including organic farming, agroforestry, integrated farming, sustainable development, and applied ecology. Permaculture has been applied most commonly to the design of housing and landscaping, integrating techniques such as agroforestry, natural building, and rainwater harvesting within the context of permaculture design principles and theory.

Theory

Twelve Design Principles

Twelve Permaculture design principles articulated by David Holmgren in his Permaculture: Principles and Pathways Beyond Sustainability:

1. Observe and interact: By taking time to engage with nature we can design solutions that suit our particular situation.

2. Catch and store energy: By developing systems that collect resources at peak abundance, we can use them in times of need.

3. Obtain a yield: Ensure that you are getting truly useful rewards as part of the work that you are doing.

4. Apply self-regulation and accept feedback: We need to discourage inappropriate activity to ensure that systems can continue to function well.

5. Use and value renewable resources and services: Make the best use of nature's abundance to reduce our consumptive behavior and dependence on non-renewable resources.

6. Produce no waste: By valuing and making use of all the resources that are available to us, nothing goes to waste.

7. Design from patterns to details: By stepping back, we can observe patterns in nature and society. These can form the backbone of our designs, with the details filled in as we go.

8. Integrate rather than segregate: By putting the right things in the right place, relationships develop between those things and they work together to support each other.

9. Use small and slow solutions: Small and slow systems are easier to maintain than big ones, making better use of local resources and producing more sustainable outcomes.

10. Use and value diversity: Diversity reduces vulnerability to a variety of threats and takes advantage of the unique nature of the environment in which it resides.

11. Use edges and value the marginal: The interface between things is where the most interesting events take place. These are often the most valuable, diverse and productive elements in the system.

12. Creatively use and respond to change: We can have a positive impact on inevitable change by carefully observing, and then intervening at the right time.

Layers

Suburban permaculture garden in Sheffield, UK with different layers of vegetation

Layers are one of the tools used to design functional ecosystems that are both sustainable and of direct benefit to humans. A mature ecosystem has a huge number of relationships between its component parts: trees, understory, ground cover, soil, fungi, insects, and animals. Because plants grow to different heights, a diverse community of life is able to grow in a relatively small space, as the vegetation occupies different layers. There are generally seven recognized layers in a food forest, although some practitioners also include fungi as an eighth layer.

1. The canopy: the tallest trees in the system. Large trees dominate but typically do not saturate the area, i.e. there exist patches barren of trees.

2. Understory layer: trees that revel in the dappled light under the canopy.

3. Shrub layer: a diverse layer of woody perennials of limited height. includes most berry bushes.

4. Herbaceous layer: Plants in this layer die back to the ground every winter (if winters are cold enough, that is). They do not produce woody stems as the Shrub layer does. Many culinary and medicinal herbs are in this layer. A large variety of beneficial plants fall into this layer. May be annuals, biennials or perennials.

5. Soil surface/Groundcover: There is some overlap with the Herbaceous layer and the Groundcover layer; however plants in this layer grow much closer to the ground, grow densely to fill bare patches of soil, and often can tolerate some foot traffic. Cover crops retain soil and lessen erosion, along with green manures that add nutrients and organic matter to the soil, especially nitrogen.

6. Rhizosphere: Root layers within the soil. The major components of this layer are the soil and the organisms that live within it such as plant roots (including root crops such as potatoes and other edible tubers), fungi, insects, nematodes, worms, etc.

7. Vertical layer: climbers or vines, such as runner beans and lima beans (vine varieties).

Guilds

There are many forms of guilds, including guilds of plants with similar functions (that could interchange within an ecosystem), but the most common perception is that of a mutual support guild. Such a guild is a group of species where each provides a unique set of diverse functions that work in conjunction, or harmony. Mutual support guilds are groups of plants, animals, insects, etc. that work well together. Some plants may be grown for food production, some have tap roots that draw nutrients up from deep in the soil, some are nitrogen-fixing legumes, some attract beneficial insects, and others repel harmful insects. When grouped together in a mutually beneficial arrangement, these plants form a guild.

Edge Effect

The edge effect in ecology is the effect of the juxtaposition or placing side by side of contrasting environments on an ecosystem. Permaculturists argue that, where vastly differing systems meet, there is an intense area of productivity and useful connections. An example of this is the coast; where the land and the sea meet there is a particularly rich area that meets a disproportionate percentage of human and animal needs. So this idea is played out in permacultural designs by using spirals in the herb garden or creating ponds that have wavy undulating shorelines rather than a simple circle or oval (thereby increasing the amount of edge for a given area).

Zones

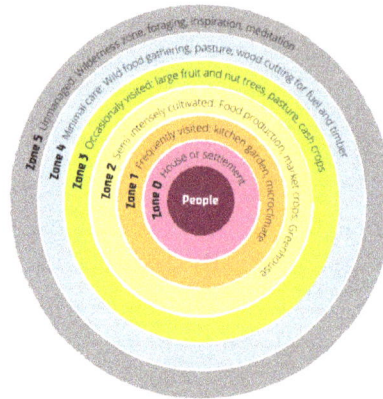

Permaculture Zones 0-5.

Zones are a way of intelligently organizing design elements in a human environment on the basis of the frequency of human use and plant or animal needs. Frequently manipulated or harvested elements of the design are located close to the house in zones 1 and 2. Less frequently used or manipulated elements, and elements that benefit from isolation (such as wild species) are farther away. Zones are about positioning things appropriately, and are numbered from 0 to 5.

Zone 0

> The house, or home center. Here permaculture principles would be applied in terms of aiming to reduce energy and water needs, harnessing natural resources such as sunlight, and generally creating a harmonious, sustainable environment in which to live and work. Zone 0 is an informal designation, which is not specifically defined in Bill Mollison's book.

Zone 1

> The zone nearest to the house, the location for those elements in the system that require frequent attention, or that need to be visited often, such as salad crops, herb plants, soft fruit like strawberries or raspberries, greenhouse and cold frames, propagation area, worm compost bin for kitchen waste, etc. Raised beds are often used in zone 1 in urban areas.

Zone 2

> This area is used for siting perennial plants that require less frequent maintenance, such as occasional weed control or pruning, including currant bushes and orchards, pumpkins, sweet potato, etc. This would also be a good place for beehives, larger scale composting bins, etc.

Zone 3

> The area where main-crops are grown, both for domestic use and for trade purposes. After establishment, care and maintenance required are fairly minimal (provided mulches and similar things are used), such as watering or weed control maybe once a week.

Zone 4

> A semi-wild area. This zone is mainly used for forage and collecting wild food as well as production of timber for construction or firewood.

Zone 5

> A wilderness area. There is no human intervention in zone 5 apart from the observation of natural ecosystems and cycles. Through this zone we build up a natural reserve of bacteria, moulds and insects that can aid the zones above it.

People and Permaculture

Permaculture uses observation of nature to create regenerative systems, and the place where this has been most visible has been on the landscape. There has been a growing awareness though that firstly, there is the need to pay more attention to the peoplecare ethic, as it is often the dynamics of people that can interfere with projects, and secondly that the principles of permaculture can be used as effectively to create vibrant, healthy and productive people and communities as they have been in landscapes.

Domesticated Animals

Domesticated animals are often incorporated into site design.

Common Practices

Agroforestry

Agroforestry is an integrated approach of using the interactive benefits from combining trees and shrubs with crops and/or livestock. It combines agricultural and forestry technologies to create more diverse, productive, profitable, healthy and sustainable land-use systems. In agroforestry systems, trees or shrubs are intentionally used within agricultural systems, or non-timber forest products are cultured in forest settings.

Forest gardening is a term permaculturalists use to describe systems designed to mimic natural forests. Forest gardens, like other permaculture designs, incorporate processes and relationships that the designers understand to be valuable in natural ecosystems. The terms forest garden and food forest are used interchangeably in the permaculture literature. Numerous permaculturists are proponents of forest gardens, such as

Graham Bell, Patrick Whitefield, Dave Jacke, Eric Toensmeier and Geoff Lawton. Bell started building his forest garden in 1991 and wrote the book The Permaculture Garden in 1995, Whitefield wrote the book How to Make a Forest Garden in 2002, Jacke and Toensmeier co-authored the two volume book set Edible Forest Gardening in 2005, and Lawton presented the film Establishing a Food Forest in 2008.

Tree Gardens, such as Kandyan tree gardens, in South and Southeast Asia, are often hundreds of years old. Whether they derived initially from experiences of cultivation and forestry, as is the case in agroforestry, or whether they derived from an understanding of forest ecosystems, as is the case for permaculture systems, is not self-evident. Many studies of these systems, especially those that predate the term permaculture, consider these systems to be forms of agroforestry. Permaculturalists who include existing and ancient systems of polycropping with woody species as examples of food forests may obscure the distinction between permaculture and agroforestry.

Food forests and agroforestry are parallel approaches that sometimes lead to similar designs.

Hügelkultur

Hügelkultur is the practice of burying large volumes of wood to increase soil water retention. The porous structure of wood acts as a sponge when decomposing underground. During the rainy season, masses of buried wood can absorb enough water to sustain crops through the dry season. This technique has been used by permaculturalists Sepp Holzer, Toby Hemenway, Paul Wheaton, and Masanobu Fukuoka.

Natural Building

A natural building involves a range of building systems and materials that place major emphasis on sustainability. Ways of achieving sustainability through natural building focus on durability and the use of minimally processed, plentiful or renewable resources, as well as those that, while recycled or salvaged, produce healthy living environments and maintain indoor air quality.

The basis of natural building is the need to lessen the environmental impact of buildings and other supporting systems, without sacrificing comfort, health, or aesthetics. To be more sustainable, natural building uses primarily abundantly available, renewable, reused, or recycled materials. In addition to relying on natural building materials, the emphasis on the architectural design is heightened. The orientation of a building, the utilization of local climate and site conditions, the emphasis on natural ventilation through design, fundamentally lessen operational costs and positively impact the environment. Building compactly and minimizing the ecological footprint is common, as are on-site handling of energy acquisition, on-site water capture, alternate sewage treatment, and water reuse.

Rainwater Harvesting

Rainwater harvesting is the accumulating and storing of rainwater for reuse before it reaches the aquifer. It has been used to provide drinking water, water for livestock, water for irrigation, as well as other typical uses. Rainwater collected from the roofs of houses and local institutions can make an important contribution to the availability of drinking water. It can supplement the subsoil water level and increase urban greenery. Water collected from the ground, sometimes from areas which are especially prepared for this purpose, is called stormwater harvesting.

Greywater is wastewater generated from domestic activities such as laundry, dishwashing, and bathing, which can be recycled on-site for uses such as landscape irrigation and constructed wetlands. Greywater is largely sterile, but not potable (drinkable). Greywater differs from water from the toilets, which is designated sewage or blackwater to indicate it contains human waste. Blackwater is septic or otherwise toxic and cannot easily be reused. There are, however, continuing efforts to make use of blackwater or human waste. The most notable is for composting through a process known as humanure; a combination of the words human and manure. Additionally, the methane in humanure can be collected and used similar to natural gas as a fuel, such as for heating or cooking, and is commonly referred to as biogas. Biogas can be harvested from the human waste and the remainder still used as humanure. Some of the simplest forms of humanure use include a composting toilet or an outhouse or dry bog surrounded by trees that are heavy feeders which can be coppiced for wood fuel. This process eliminates the use of a standard toilet with plumbing.

Sheet Mulching

In agriculture and gardening, mulch is a protective cover placed over the soil. Any material or combination can be used as mulch, such as stones, leaves, cardboard, wood chips, gravel, etc., though in permaculture mulches of organic material are the most common because they perform more functions. These include absorbing rainfall, reducing evaporation, providing nutrients, increasing organic matter in the soil, feeding and creating habitat for soil organisms, suppressing weed growth and seed germination, moderating diurnal temperature swings, protecting against frost, and reducing erosion. Sheet mulching is an agricultural no-dig gardening technique that attempts to mimic natural processes occurring within forests. Sheet mulching mimics the leaf cover that is found on forest floors. When deployed properly and in combination with other Permacultural principles, it can generate healthy, productive and low maintenance ecosystems.

Sheet mulch serves as a "nutrient bank," storing the nutrients contained in organic matter and slowly making these nutrients available to plants as the organic matter slowly and naturally breaks down. It also improves the soil by attracting and feeding earthworms, slaters and many other soil micro-organisms, as well as adding humus.

Earthworms "till" the soil, and their worm castings are among the best fertilizers and soil conditioners. Sheet mulching can be used to reduce or eliminate undesirable plants by starving them of light, and can be more advantageous than using herbicide or other methods of control.

Intensive Rotational Grazing

Grazing has long been blamed for much of the destruction we see in the environment. However, it has been shown that when grazing is modeled after nature, the opposite effect can be seen. Also known as cell grazing, managed intensive rotational grazing (MIRG) is a system of grazing in which ruminant and non-ruminant herds and/or flocks are regularly and systematically moved to fresh pasture, range, or forest with the intent to maximize the quality and quantity of forage growth. This disturbance is then followed by a period of rest which allows new growth. MIRG can be used with cattle, sheep, goats, pigs, chickens, rabbits, geese, turkeys, ducks, and other animals depending on the natural ecological community that is being mimicked. Sepp Holzer and Joel Salatin have shown how the disturbance caused by the animals can be the spark needed to start ecological succession or prepare ground for planting. Allan Savory's holistic management technique has been likened to "a permaculture approach to rangeland management". One variation on MIRG that is gaining rapid popularity is called eco-grazing. Often used to either control invasives or re-establish native species, in eco-grazing the primary purpose of the animals is to benefit the environment and the animals can be, but are not necessarily, used for meat, milk or fiber.

Keyline Design

Keyline design is a technique for maximizing beneficial use of water resources of a piece of land developed in Australia by farmer and engineer P. A. Yeomans. The Keyline refers to a specific topographic feature linked to water flow which is used in designing the drainage system of the site.

Fruit Tree Management

Some proponents of permaculture advocate no, or limited, pruning. One advocate of this approach is Sepp Holzer who used the method in connection with Hügelkultur berms. He has successfully grown several varieties of fruiting trees at altitudes (approximately 9,000 feet (2,700 m)) far above their normal altitude, temperature, and snow load ranges. He notes that the Hügelkultur berms kept and/or generated enough heat to allow the roots to survive during alpine winter conditions. The point of having unpruned branches, he notes, was that the longer (more naturally formed) branches bend over under the snow load until they touched the ground, thus forming a natural arch against snow loads that would break a shorter, pruned, branch.

Masanobu Fukuoka, as part of early experiments on his family farm in Japan, experimented with no-pruning methods, noting that he ended up killing many fruit trees by simply letting them go, which made them become convoluted and tangled, and thus unhealthy. Then he realised this is the difference between natural-form fruit trees and the process of change of tree form that results from abandoning previously-pruned unnatural fruit trees. He concluded that the trees should be raised all their lives without pruning, so they form healthy and efficient branch patterns that follow their natural inclination. This is part of his implementation of the Tao-philosophy of Wú wéi translated in part as no-action (against nature), and he described it as no unnecessary pruning, nature farming or "do-nothing" farming, of fruit trees, distinct from non-intervention or literal no-pruning. He ultimately achieved yields comparable to or exceeding standard/intensive practices of using pruning and chemical fertilisation.

Trademark and Copyright Issues

There has been contention over who, if anyone, controls legal rights to the word permaculture: is it trademarked or copyrighted? and if so, who holds the legal rights to the use of the word? For a long time Bill Mollison claimed to have copyrighted the word, and his books said on the copyright page, "The contents of this book and the word PERMACULTURE are copyright." These statements were largely accepted at face-value within the permaculture community. However, copyright law does not protect names, ideas, concepts, systems, or methods of doing something; it only protects the expression or the description of an idea, not the idea itself. Eventually Mollison acknowledged that he was mistaken and that no copyright protection existed for the word permaculture.

In 2000, Mollison's US based Permaculture Institute sought a service mark (a form of trademark) for the word permaculture when used in educational services such as conducting classes, seminars, or workshops. The service mark would have allowed Mollison and his two Permaculture Institutes (one in the US and one in Australia) to set enforceable guidelines regarding how permaculture could be taught and who could teach it, particularly with relation to the PDC, despite the fact that he had instituted a system of certification of teachers to teach the PDC in 1993. The service mark failed and was abandoned in 2001. Also in 2001 Mollison applied for trademarks in Australia for the terms "Permaculture Design Course" and "Permaculture Design". These applications were both withdrawn in 2003. In 2009 he sought a trademark for "Permaculture: A Designers' Manual" and "Introduction to Permaculture", the names of two of his books. These applications were withdrawn in 2011. There has never been a trademark for the word permaculture in Australia.

Criticisms

General Criticisms

In 2011, Owen Hablutzel argued that "permaculture has yet to gain a large amount of specific mainstream scientific acceptance," and that "the sensitiveness to being perceived and accepted on scientific terms is motivated in part by a desire for permaculture to expand and become increasingly relevant."

In his books Sustainable Freshwater Aquaculture and Farming in Ponds and Dams, Nick Romanowski expresses the view that the presentation of aquaculture in Bill Mollison's books is unrealistic and misleading.

Agroforestry

Greg Williams argues that forests cannot be more productive than farmland because the net productivity of forests decline as they mature due to ecological succession. Proponents of permaculture respond that this is true only if one compares data between woodland forest and climax vegetation, but not when comparing farmland vegetation with woodland forest. For example, ecological succession generally results in a forest's productivity rising after its establishment only until it reaches the woodland state (67% tree cover), before declining until full maturity.

Vegan Organic Gardening

Vegan organic gardening and farming is the organic cultivation and production of food crops and other crops with a minimal amount of exploitation or harm to any animal. Vegan gardening and stock-free farming methods use no animal products or by-products, such as bloodmeal, fish products, bone meal, feces, or other animal-origin matter, because the production of these materials is viewed as either harming animals directly, or being associated with the exploitation and consequent suffering of animals. Some of these materials are by-products of animal husbandry, created during the process of cultivating animals for the production of meat, milk, skins, furs, entertainment, labor, or companionship; the sale of by-products decreases expenses and increases profit for those engaged in animal husbandry, and therefore helps support the animal husbandry industry, an outcome most vegans find unacceptable.

Types

Veganiculture

Vegan - Organic - Agriculture / Permaculture The Future Of Farming! All Things Related To: Organic Gardening, Farming & Food Forests Free From Animals & Animal Products

Forest Gardening

Forest gardening is a fully plant-based organic food production system based on woodland ecosystems, incorporating fruit and nut trees, shrubs, herbs, vines and perennial vegetables. Making use of companion planting, these can be intermixed to grow in a succession of layers, to replicate a woodland habitat. Forest gardening can be viewed as a way to recreate the Garden of Eden. The three main products from a forest garden are fruit, nuts and green leafy vegetables.

Robert Hart's forest garden in Shropshire, England.

Robert Hart adapted forest gardening for temperate zones during the early 1960s. Robert Hart began with a conventional smallholding at Wenlock Edge in Shropshire. However, following his adoption of a raw vegan diet for health and personal reasons, Hart replaced his farm animals with plants. He created a model forest garden from a small orchard on his farm and intended naming his gardening method ecological horticulture or ecocultivation. Hart later dropped these terms once he became aware that agroforestry and forest gardens were already being used to describe similar systems in other parts of the world.

Vegan Permaculture

Vegan permaculture (also known as veganic permaculture, veganiculture, or vegaculture) avoids the use of domesticated animals. It is essentially the same as permaculture except for the addition of a fourth core value; "Animal Care." Zalan Glen, a raw vegan, proposes that vegaculture should emerge from permaculture in the same way veganism split from vegetarianism in the 1940s. Vegan permaculture recognizes the importance of free-living animals, not domesticated animals, to create a balanced ecosystem.

Veganic Gardening

The veganic gardening method is a distinct system developed by Rosa Dalziell O'Brien, Kenneth Dalziel O'Brien and May E. Bruce, although the term was originally coined by Geoffrey Rudd as a contraction of vegetable organic in order to "denote a clear distinction between conventional chemical based systems and organic ones based on animal manures". The O'Brien system's principal argument is that animal manures are harmful to soil health rather than that their use involves exploitation of and cruelty to animals.

The system employs very specific techniques including the addition of straw and other vegetable wastes to the soil in order to maintain soil fertility. Gardeners following the system use soil-covering mulches, and employ non-compacting surface cultivation techniques using any short-handled, wide-bladed, hand hoe. They kneel when surface cultivating, placing a board under their knees to spread out the pressure, and prevent soil compaction. Kenneth Dalziel O'Brien published a description of his system in Veganic Gardening, the Alternative System for Healthier Crops:

The veganic method of clearing heavily infested land is to take advantage of a plant's tendencies to move its roots nearer to the soil's surface when it is deprived of light. To make use of this principle, aided by a decaying process of the top growth of weeds, etc., it is necessary to subject such growth to heat and moisture in order to speed up the decay, and this is done by applying lime, then a heavy straw cover, and then the herbal compost activator...The following are required: Sufficient new straw to cover an area to be cleared to a depth of 3 to 4 inches.

The O'Brien method also advocates minimal disturbance of the soil by tilling, the use of cover crops and green manures, the creation of permanent raised beds and permanent hard-packed paths between them, the alignment of beds along a north-south axis, and planting in double rows or more so that not every row has a path on both sides. Use of animal manure is prohibited.

Vegan Biodynamic Agriculture

The German agricultural researcher Maria Thun (1922 - 2012) developed vegan equivalents to the traditional, animal based biodynamic preparations. As a reaction to the BSE scandal in Europe she started researching plant based preparations, using tree barks as replacement for animal organs as sheath for the preparations.

In particular in Italy, there is a movement of vegan biodynamic farming, represented by farmers such as Sebastiano Cossia Castiglioni and Cristina Menicocci.

There are many other methods currently used and under development.

Practices

Soil fertility is maintained by the use of green manures, cover crops, green wastes, composted vegetable matter, and minerals. Some vegan gardeners may supplement this with human urine from vegans (which provides nitrogen) and 'humanure' from vegans, produced from compost toilets. Generally only waste from vegans is used because of the expert recommendation that the risks associated with using composted waste are acceptable only if the waste is from animals or humans having a largely herbivorous diet.

Veganic gardeners may prepare soil for cultivation using the same method used by conventional and organic gardeners of breaking up the soil with hand tools and power tools and allowing the weeds to decompose.

Xeriscaping

The Xeriscape Demonstration Garden at the headquarters of Denver Water in Denver, Colorado.

Xeriscaping is landscaping and gardening that reduces or eliminates the need for supplemental water from irrigation. It is promoted in regions that do not have easily accessible, plentiful, or reliable supplies of fresh water, and is gaining acceptance in other areas as access to water becomes more limited. Xeriscaping may be an alternative to various types of traditional gardening.

In some areas, terms as water-conserving landscapes, drought-tolerant landscaping, and smart scaping are used instead. Plants whose natural requirements are appropriate to the local climate are emphasized, and care is taken to avoid losing water to evaporation and run-off. The specific plants used in xeriscaping depend upon the climate. Xeriscaping is different from natural landscaping, because the emphasis in xeriscaping is on selection of plants for water conservation, not necessarily selecting native plants.

Public perception of xeriscaping has frequently been negative as many assume that these types of landscapes are ugly expanses of cactus and gravel. However, studies have shown that education in water conservation practices in the landscape can greatly improve the public's perception of xeriscaping.

Similar Terms

The term zero-scaping or zeroscaping is sometimes substituted for xeriscaping due to phonetic similarity. When used seriously, zero-scaping usually refers to a different type of low-water landscaping that is often devoid, or nearly devoid of plants. Because the term was derived from the Greek root xeros, xeriscaping is sometimes misspelled xeroscaping.

Advantages

Cacti are one of the low-water-consuming plants used in Xeriscaping.

- Lowered consumption of water: Xeriscaped landscapes use up to two thirds less water than regular lawn landscapes.

- Makes more water available for other domestic and community uses and the environment.

- Reduce Maintenance: Aside from occasional weeding and mulching Xeriscaping requires far less time and effort to maintain.

- Xeriscape plants in appropriate planting design, and soil grading and mulching, takes full advantage of rainfall retention.

- Less cost to maintain: Xeriscaping requires less fertilisers and equipment, particularly due to the reduced lawn areas.

- Reduced waste and pollution: Lawn clippings can contribute to organic waste in landfills and the use of heavy fertilisers contributes to urban runoff pollution.

Criticisms and Refutations

- Xeriscape style may not conform to local aesthetics: Some homeowners associations have strict rules requiring a certain percentage of land to be used as lawns. Refutation: Because of drought and improved education, these rules either have been or are in the process of being overturned in many areas.

- Xeriscape contains hazardous plants: Some styles of xeriscaping include plants such as cacti and agave having thorns or serrated edges that may harm pets and people. Refutation: There are many xeric plants that possess no sharp points, and many familiar plants like rose bushes and raspberry plants have thorns.

- Xeriscape installation cost is high: If a non-xeric landscape was originally present, the cost of removing the existing landscape and installing a xeric one may be a deterrent. Refutation: For new construction, xeric landscaping can cost much less than a lawn. Whether new or a replacement, the water-saving cost benefit of a xeric landscape over a non-xeric one is self-evident. Over time the cost savings can pay for the installation. In arid and semi-arid areas, drought-based water restrictions may necessitate the reduction of turf areas that would be likely to die due to a lack of water already occurring because of the drought.

Principles

The Al Norris Memorial Xeriscape Garden in Wichita Falls, Texas

Originally conceived by Denver Water, the seven design principles of xeriscaping have since expanded into simple and applicable concepts to creating landscapes that use less water. The principles are appropriate for multiple regions and can serve as a guide to creating a water conserving landscape that is regionally appropriate and since they were conceived for homeowners they are easy to implement.

1. Plan and design: Create a diagram, drawn to scale, that shows the major elements of the landscape, including house, driveway, sidewalk, deck or patio, existing trees and other elements.

Once a base plan of an existing site has been determined, the creation of a conceptual plan (bubble diagram) that shows the areas for turf, perennial beds, views, screens, slopes, etc. is undertaken. Once finished, the development of a planting plan that reinforces the areas in the appropriate scale is done.

2. Soil amendment – Most plants will benefit from the use of compost, which will help the soil retain water. Some desert plants prefer gravel soils instead of well-amended soils. Plants should either fit the soil or soil should be amended to fit the plants.

3. Efficient irrigation: Xeriscape can be irrigated efficiently by hand or with an automatic sprinkler system. Zone turf areas separately from other plants and use the irrigation method that waters the plants in each area most efficiently. For grass, use gear-driven

rotors or rotary spray nozzles that have larger droplets and low angles to avoid wind drift. Spray, drip line or bubbler emitters are most efficient for watering trees, shrubs, flowers and groundcovers.

If watering by hand, avoid oscillating sprinklers and other sprinklers that throw water high in the air or release a fine mist. The most efficient sprinklers release big drops close to the ground.

Water deeply and infrequently to develop deep roots. Never water during the day to reduce water lost to evaporation. With the use of automatic sprinkling systems, adjust the controller monthly to accommodate weather conditions. Also, install a rain sensor to shut off the device when it rains.

4. Appropriate plant and zone selection: Different areas in a yard receive different amounts of light, wind and moisture. To minimize water waste, group together plants with similar light and water requirements, and place them in an area that matches these requirements. Put moderate-water-use plants in low-lying drainage areas, near downspouts, or in the shade of other plants. Turf typically requires the most water and shrub/perennial beds will require approximately half the amount of water. Dry, sunny areas support low-water-use plants that grow well in our climate. Planting a variety of plants with different heights, color and textures creates interest and beauty.

5. Mulch: Mulch keeps plant roots cool, prevents soil from crusting, minimizes evaporation and reduces weed growth. Organic mulches, such as bark chips, pole peelings or wood grindings, should be applied 2 to 4 inches deep. Fiber mulches create a web that is more resistant to wind and rain washout. Inorganic mulches, such as rocks and gravel, should be applied 2 to 3 inches deep. Surrounding plants with rock makes the area hotter; limit this practice.

6. Limited turf areas: Native grasses (warm-season) that have been cultivated for turf lawns, such as buffalo grass and blue grama, can survive with a quarter of the water that bluegrass varieties need. Warm-season grasses are greenest in June through September and straw brown the rest of the year.

Native grasses (cool season) such as bluegrass and tall fescue, are greenest in the spring and fall and go dormant in the high heat of the summer. New cultivars of bluegrass, such as Reveille, and tall fescue, can reduce typical bluegrass water requirements by at least 30 percent. Fine fescues can provide substantial water savings and is best used in areas that receive low traffic or are in shady locations.

Use the appropriate grass and limit the amount of grass to reduce the watering and maintenance requirements.

7. Maintenance: All landscapes require some degree of care during the year. Turf requires spring and fall aeration along with regular fertilization every 6 to 8 weeks. Keep the grass height at 3 inches and allow the clippings to fall. Trees, shrubs and perenni-

als will need occasional pruning to remove dead stems, promote blooming or control height and spread. Much of the removed plant material can be shredded and used in composting piles.

Lawns

One of the major challenges to the public acceptance of xeriscaping is the cultural attachment to turf grass lawns. Originally implemented in England, lawns have become a universal symbol of prosperity, order and community. In the United States turf grasses are so common that it is the single most irrigated crop by surface area, covering nearly 128,000 square kilometres (49,000 sq mi). Despite the high water, fertiliser and maintenance costs associated with lawns, they have become the norm in most urban and suburban areas, even if they are rarely used for recreational purposes or otherwise. Xeriscaped landscapes offer an alternative to the over use of turf grass lawns but are not widely accepted because of preconceived notions of what it means to xeriscape. Xeriscaping can include lawn areas but seeks to reduce them to areas that will actually be used, rather than using them as a default landscaping plan.

Wildlife Garden

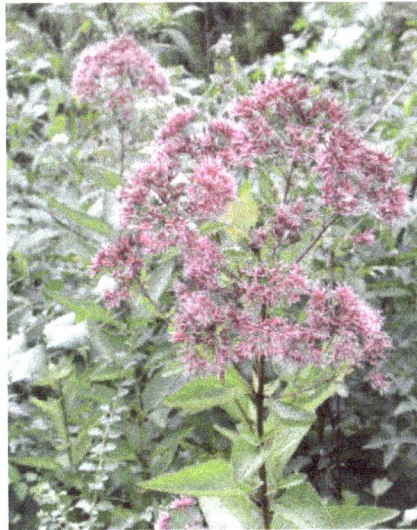

Joe-Pye weed in flower

A wildlife garden (or wild garden) is an environment created by a gardener that serves as a sustainable haven for surrounding wildlife. Wildlife gardens contain a variety of habitats that cater to native and local plants, birds, amphibians, reptiles, insects, mammals and so on. Establishing a garden environment that mimics surrounding wildlife allows for natural systems to interact and establish an equilibrium, ultimately minimizing the need for gardener maintenance and intervention. Wildlife gardens can also play an essential role in biological pest control, and also promote biodiversity, native plantings, and generally benefit the wider environment.

Habitats

Building a successful garden suitable for local wildlife is best accomplished through the use of multiple three-dimensional habitats with diverse structures that provide places for animals to nest and hide. Wildlife gardens may contain a range of habitats, including:

Log piles – Preferably located in a shady area, a pile of logs is a sanctuary for insects and amphibians. The organic structure is a shelter for both protection and breeding. In addition to logs, garden debris may also be added around the garden to be used as a natural mulch, fertilizer, weed control, soil amendment, and habitat for arthropod predators.

Bird feeding stations and bird houses – A place for birds to eat and take shelter will increase the amount of birds in the garden, which play a key role in biological pest control. Not only will food and shelter increase the survival rate of birds, but it will also ensure that they are healthy enough for a successful breeding season.

Bug boxes – Offcuts of wood placed in a structure above ground provides an alternate place of shelter for beneficial insects, such as the robber fly, which help keep natural ecosystem predators in check.

Sources of water – A water feature, such as a pond, has the potential to support a large biodiversity of wildlife. To maximize the amount of wildlife attracted to the water feature, it should consist of ranging depths. Shallow areas are used by birds to drink and by insects and amphibians to lay eggs. Deeper areas provide habitat for aquatic insects and a place for amphibians, or even fish to swim.

Pollinating flowers – Flowers rich in nectar will attract bees and butterflies into the garden. Wildflower meadows are an alternative option for lawns in the garden and will serve as a sanctuary for pollinators. However, pollinating plants should not be confused with plants suitable for butterfly breeding.

Plant diversity – The garden should include a range of plant types to serve different specie habitats. A balance between ground cover, shrub, understory, and canopy species will allow different sized wildlife shelters that fit their individual needs.

Choice of Plants

Although some exotics may also be included, wild gardens usually mostly feature a variety of native species. Generally, these will be a part of the pre-existing natural ecology of an area, making them easier to grow than most exotic species. Choosing native plants comes with an array of benefits for both plant and animal diversity, especially the ability to support native insect and mushroom populations that have established balanced evolutionary relationships over thousands of years.

Ornamental plants on the market tend to lean toward "pest-free" plants, making it hard for native insects to adapt, and ultimately reducing their food supply. Decreases in insect populations due to excessive ornamental planting will discourage bird populations from inhabiting the area.

Invasive species can always prove problematic in the garden due to the absence of natural predators and their ability to reproduce rapidly. Without any measures of control, invasive species can easily overtake native species in the garden. Addressing invasive plants can be done a variety of ways; however, to ensure the least amount of damage to the surrounding ecosystem, this is best done by cutting down the plant, followed by painting its stem with an herbicide, such as Roundup. The debris from the invasive species can be piled and used as a home for smaller critters.

Wildlife Gardens in the Netherlands

Wildlife gardens in the Netherlands are called "heemtuinen". The first was created in 1925: Thijsse's Hof (Garden of Thijsse) in Bloemendaal, near Haarlem. It was gifted to Jac. P. Thijsse on the occasion of his 60th anniversary, and still exists today. The garden gives a display of about 800 plants native to the dune region of South Kennemerland, in which the garden is situated. It is said to be one of the oldest wildlife gardens of its sort in the world.

Nowadays some 25 wildlife gardens exist in the Netherlands.

Forest Gardening

Robert Hart's forest garden in Shropshire

Forest gardening is a low-maintenance sustainable plant-based food production and agroforestry system based on woodland ecosystems, incorporating fruit and nut trees, shrubs, herbs, vines and perennial vegetables which have yields directly useful to humans. Making use of companion planting, these can be intermixed to grow in a succession of layers, to build a woodland habitat.

Forest gardening is a prehistoric method of securing food in tropical areas. In the 1980s, Robert Hart coined the term "forest gardening" after adapting the principles and applying them to temperate climates.

History

Forest gardens are probably the world's oldest form of land use and most resilient agro-ecosystem. They originated in prehistoric times along jungle-clad river banks and in the wet foothills of monsoon regions. In the gradual process of families improving their immediate environment, useful tree and vine species were identified, protected and improved whilst undesirable species were eliminated. Eventually superior foreign species were selected and incorporated into the gardens.

Forest gardens are still common in the tropics and known by various names such as: home gardens in Kerala in South India, Nepal, Zambia, Zimbabwe and Tanzania; Kandyan forest gardens in Sri Lanka; huertos familiares, the "family orchards" of Mexico; and pekarangan, the gardens of "complete design", in Java. These are also called agroforests and, where the wood components are short-statured, the term shrub garden is employed. Forest gardens have been shown to be a significant source of income and food security for local populations.

Robert Hart adapted forest gardening for the United Kingdom's temperate climate during the 1980s. His theories were later developed by Martin Crawford from the Agroforestry Research Trust and various permaculturalists such as Graham Bell, Patrick Whitefield, Dave Jacke and Geoff Lawton.

In Tropical Climates

Forest gardens, or home gardens, are common in the tropics, using intercropping to cultivate trees, crops, and livestock on the same land. In Kerala in south India as well as in northeastern India, the home garden is the most common form of land use and is also found in Indonesia. One example combines coconut, black pepper, cocoa and pineapple. These gardens exemplify polyculture, and conserve much crop genetic diversity and heirloom plants that are not found in monocultures. Forest gardens have been loosely compared to the religious concept of the Garden of Eden.

Americas

The BBC's Unnatural Histories claimed that the Amazon rainforest, rather than being a pristine wilderness, has been shaped by humans for at least 11,000 years through prac- tices such as forest gardening and terra preta. This was also explored in the bestselling book 1491 by author Charles C. Mann. Since the 1970s, numerous geoglyphs have also been discovered on deforested land in the Amazon rainforest, furthering the evidence about Pre-Columbian civilizations.

On the Yucatán Peninsula, much of the Maya food supply was grown in "orchard-gar- dens", known as pet kot. The system takes its name from the low wall of stones (pet meaning circular and kot wall of loose stones) that characteristically surrounds the gardens.

Africa

In many African countries, for example Zambia, Zimbabwe, Ethiopia and Tanzania, gardens are widespread in rural, periurban and urban areas and they play an essential role in establishing food security. Most well known are the Chaga or Chagga gardens on the slopes of Mt. Kilimanjaro in Tanzania. These are an excellent example of an agro- forestry system. In many countries, women are the main actors in home gardening and food is mainly produced for subsistence. In North-Africa, oasis layered gardening with palm trees, fruit trees and vegetables is a traditional type of forest garden.

Nepal

In Nepal, the Ghar Bagaincha, literally "home garden", refers to the traditional land use system around a homestead, where several species of plants are grown and main- tained by household members and their products are primarily intended for the fam- ily consumption (Shrestha et al., 2002). The term "home garden" is often considered synonymous to the kitchen garden. However, they differ in terms of function, size, di- versity, composition and features (Sthapit et al., 2006). In Nepal, 72% of households have home gardens of an area 2–11% of the total land holdings (Gautam et al., 2004). Because of their small size, the government has never identified home gardens as an important unit of food production and they thereby remain neglected from research and development. However, at the household level the system is very important as it is an important source of quality food and nutrition for the rural poor and, therefore, are important contributors to the household food security and livelihoods of farming com- munities in Nepal. The gardens are typically cultivated with a mixture of annual and perennial plants that can be harvested on a daily or seasonal basis. Biodiversity that has an immediate value is maintained in home gardens as women and children have easy access to preferred food. Home gardens, with their intensive and multiple uses, provide a safety net for households when food is scarce. These gardens are not only important sources of food, fodder, fuel, medicines, spices, herbs, flowers, construction materials and income in many countries, they are also important for the in situ con- servation of a wide range of unique genetic resources for food and agriculture (Subedi et al., 2004). Many uncultivated, as well as neglected and underutilised species could make an important contribution to the dietary diversity of local communities (Gautam et al., 2004).

In addition to supplementing diet in times of difficulty, home gardens promote whole-family and whole-community involvement in the process of providing food. Children, the elderly, and those caring for them can participate in this infield agriculture, incorporating it with other household tasks and scheduling. This tradition has existed in many cultures around the world for thousands of years.

In Mediterranean Climates

The Mediterranean climate has long, hot, rainless summers and relatively short, cool, rainy winters (Köppen climate classification Csa). Its climate conditions are highly variable within an area and modified locally by altitude, latitude, and the proximity to the Mediterranean. In the 1950s the Forest Research Department of the Ministry of Agriculture founded a botanical forest garden in the Sharon region in Israel, the Ilanot Forest. As the only one of its kind in Israel, it harbours more than 750 species of trees from locations all over the world, including the Japanese sago palm cycas revoluta, fig trees (ficus glomerata), stone pine trees (pinus pinea) that produce tasty pine nuts and adds to the biodiversity of Israel.

In Temperate Climates

Robert Hart, forest gardening pioneer

Robert Hart coined the term "forest gardening" during the 1980s. Hart began farming at Wenlock Edge in Shropshire with the intention of providing a healthy and therapeutic environment for himself and his brother Lacon. Starting as relatively conventional smallholders, Hart soon discovered that maintaining large annual vegetable beds, rearing livestock and taking care of an orchard were tasks beyond their strength. However, a small bed of perennial vegetables and herbs he planted was looking after itself with little intervention.

Following Hart's adoption of a raw vegan diet for health and personal reasons, he replaced his farm animals with plants. The three main products from a forest garden are fruit, nuts and green leafy vegetables. He created a model forest garden from a 0.12 acre (500 m²) orchard on his farm and intended naming his gardening method ecological horticulture or ecocultivation. Hart later dropped these terms once he became aware that agroforestry and forest gardens were already being used to describe similar systems in other parts of the world. He was inspired by the forest farming methods of Toyohiko Kagawa and James Sholto Douglas, and the productivity of the Keralan home gardens as Hart explains:

From the agroforestry point of view, perhaps the world's most advanced country is the Indian state of Kerala, which boasts no fewer than three and a half million forest gardens...As an example of the extraordinary intensivity of cultivation of some forest gardens, one plot of only 0.12 hectares (0.30 acres) was found by a study group to have twenty-three young coconut palms, twelve cloves, fifty-six bananas, and forty-nine pineapples, with thirty pepper vines trained up its trees. In addition, the small holder grew fodder for his house-cow.

Seven-layer System

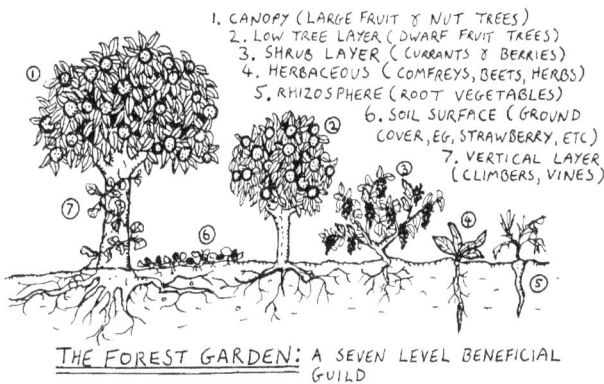

1. CANOPY (LARGE FRUIT & NUT TREES)
2. LOW TREE LAYER (DWARF FRUIT TREES)
3. SHRUB LAYER (CURRANTS & BERRIES)
4. HERBACEOUS (COMFREYS, BEETS, HERBS)
5. RHIZOSPHERE (ROOT VEGETABLES)
6. SOIL SURFACE (GROUND COVER, EG, STRAWBERRY, ETC)
7. VERTICAL LAYER (CLIMBERS, VINES)

THE FOREST GARDEN: A SEVEN LEVEL BENEFICIAL GUILD

The seven layers of the forest garden

Robert Hart pioneered a system based on the observation that the natural forest can be divided into distinct levels. He used intercropping to develop an existing small orchard of apples and pears into an edible polyculture landscape consisting of the following layers:

1. 'Canopy layer' consisting of the original mature fruit trees.

2. 'Low-tree layer' of smaller nut and fruit trees on dwarfing root stocks.

3. 'Shrub layer' of fruit bushes such as currants and berries.

4. 'Herbaceous layer' of perennial vegetables and herbs.

5. 'Rhizosphere' or 'underground' dimension of plants grown for their roots and tubers.

6. 'Ground cover layer' of edible plants that spread horizontally.

7. 'Vertical layer' of vines and climbers.

A key component of the seven-layer system was the plants he selected. Most of the traditional vegetable crops grown today, such as carrots, are sun loving plants not well selected for the more shady forest garden system. Hart favoured shade tolerant perennial vegetables.

Further Development

The Agroforestry Research Trust (ART), managed by Martin Crawford, runs experimental forest gardening projects on a number of plots in Devon, United Kingdom. Crawford describes a forest garden as a low-maintenance way of sustainably producing food and other household products.

Ken Fern had the idea that for a successful temperate forest garden a wider range of edible shade tolerant plants would need to be used. To this end, Fern created the organisation Plants for a Future (PFAF) which compiled a plant database suitable for such a system. Fern used the term woodland gardening, rather than forest gardening, in his book Plants for a Future.

The Movement for Compassionate Living (MCL) promote forest gardening and other types of vegan organic gardening to meet society's needs for food and natural resources. Kathleen Jannaway, the founder of MCL, wrote a book outlining a sustainable vegan future called Abundant Living in the Coming Age of the Tree in 1991. In 2009, the MCL provided a grant of £1,000 to the Bangor Forest Garden project in Gwynedd, North West Wales.

Kevin Bradley coined the phrase "Edible Forest" in the 1980s as the name of his nursery, garden, and orchard on 5 acres in the frigid zone 3 pine forests of northern Wisconsin. Among 3 options, he chose "Edible Forest" because it "evokes at once an ethereal, spiritual, and magical image", of Disney- like "Forest of No Return"; of the biblical "Garden of Eden". This image was perfectly in line with his ongoing experiment begun in 1985 in what he calls a closed loop human environment, combining multi- story tree and field crop "garden/orchards" for maximum beauty and use of space, someday to be very useful in an ever shrinking world. "The name, at the same time, with its irrational first impression (of course we can't eat a forest), forces the mind to think, if just a little bit, about its inference and thus sticks in our memories". It appeared from Bradley's research that the two words had, prior to the 80's, never been put together before as a noun phrase but which by today, after more than two decades of Bradley's "Edible Forest Nursery" and the 2005 text by Jacke and Toensmeirer's- "Edible Forest Gardens", has grown into a movement and little "Edible Forests" all over the world.

In 2005, Dave Jacke and Eric Toensmeier's two-volume Edible Forest Gardens provided a deeply researched reference focused on North American forest gardening climates, habitats, and species. The book attempts to ground forest gardening deeply in ecological science. The Apios Institute wiki grew out of their work, and seeks to document and share the experience of people around the world working with the species in polycultures.

Permaculture

Bill Mollison, who coined the term permaculture, visited Robert Hart at his forest garden in Wenlock Edge in October 1990. Hart's seven-layer system has since been adopted as a common permaculture design element.

Numerous permaculturalists are proponents of forest gardens, or food forests, such as Graham Bell, Patrick Whitefield, Dave Jacke, Eric Toensmeier and Geoff Lawton. Bell started building his forest garden in 1991 and wrote the book The Permaculture Garden in 1995, Whitefield wrote the book How to Make a Forest Garden in 2002, Jacke and Toensmeier co-authored the two volume book set Edible Forest Gardening in 2005, and Lawton presented the film Establishing a Food Forest in 2008.

Austrian Sepp Holzer practices "Holzer Permaculture" on his Krameterhof farm, at varying altitudes ranging from 1,100 to 1,500 metres above sea level. His designs create micro-climates with rocks, ponds and living wind barriers, enabling the cultivation of a variety of fruit trees, vegetables and flowers in a region that averages 4 °C, and with temperatures as low as -20 °C in the winter.

Projects

El Pilar on the Belize-Guatemala border features a forest garden to demonstrate traditional Maya agricultural practices. A further 1-acre model forest garden, called Känan K'aax (meaning well-tended garden in Mayan), is being funded by the National Geographic Society and developed at Santa Familia Primary School in Cayo.

In the United States the largest known food forest on public land is believed to be the 7-acre Beacon Food Forest in Seattle, Washington. Other forest garden projects include those at the Central Rocky Mountain Permaculture Institute in Basalt, Colorado and Montview Neighborhood farm in Northampton, Massachusetts.

In Canada food forester Richard Walker has been developing and maintaining food forests in the province of British Columbia for over 30 years. He developed a 3-acre food forest that when at maturity provided raw materials for a nursery and herbalism business as well as food for his family. The Living Centre have developed various forest garden projects in Ontario.

In the United Kingdom, other than those run by the Agroforestry Research Trust (ART), there are numerous forest garden projects such as the Bangor Forest Garden in Gwynedd, North West Wales. Martin Crawford from ART administers the Forest Garden Network, an informal network of people and organisations around the world who are cultivating their own forest gardens.

6

Evolution of Gardening

The history of ornamental gardening is a way of displaying beauty through art and nature. It is put on display either for private status or for national pride. The section helps the readers in understanding the evolution of gardening.

History of Gardening

The renaissance style gardens at Chateau Villandry.

The history of ornamental gardening may be considered as aesthetic expressions of beauty through art and nature, a display of taste or style in civilized life, an expression of an individual's or culture's philosophy, and sometimes as a display of private status or national pride—in private and public landscapes.

Introduction

Forest gardening, a plant-based food pro-system, is the world's oldest form of gardening. Forest gardens originated in prehistoric times along jungle-clad river banks and in the wet foothills of monsoon regions. In the gradual process of families improving their immediate environment, useful tree and vine species were identified, protected and improved whilst undesirable species were eliminated. Eventually foreign species were also selected and incorporated into the gardens.

The enclosure of outdoor space began in 10,000 BC. Though no one knows the specific details of the first garden, historians imagine the first enclosure was a type of barrier for the purpose of keeping out animals and marauders. Garden-making and design was a key precursor to landscape architecture, and it began in West Asia, eventually spreading westward into Greece, Spain, Germany, France, Britain, etc. Modern words of "garden" and "yard" are descendants of the Old English term "geard," which means fence or enclosure.

Vitruvius, a Roman author and engineer, wrote the oldest surviving design manual in 27 BC. De architecture libri decem (The Ten Books on Architecture) addressed design theory, landscape architecture, engineering, water supply and public projects, such as parks and squares. Vitruvius asserted that firmitas (firmness, durability, strength), utilitas (commodity, convenience, utility) and venustas (delight, loveliness, beauty) were the main design objectives, and some consider these elements centrally important to quality landscape design.

After the emergence of the first civilizations, wealthy individuals began to create gardens for purely aesthetic purposes. Egyptian tomb paintings of the 16th century BC are some of the earliest physical evidence of ornamental horticulture and landscape design; they depict lotus ponds surrounded by symmetrical rows of acacias and palms. Another ancient gardening tradition is of Persia: Darius the Great was said to have had a "paradise garden" and the Hanging Gardens of Babylon were renowned as a Wonder of the World. Persian gardens were also organized symmetrically, along a center line known as an axis.

Robert Hart's forest garden in Shropshire, England.

Persian influences extended to post-Alexander's Greece: around 350 BC there were gardens at the Academy of Athens, and Theophrastus, who wrote on botany, was supposed to have inherited a garden from Aristotle. Epicurus also had a garden where he walked and taught, and bequeathed it to Hermarchus of Mytilene. Alciphron also mentions private gardens.

The most influential ancient gardens in the western world were the Ptolemy's gardens at Alexandria and the gardening tradition brought to Rome by Lucullus. Wall paintings in Pompeii attest to elaborate development later. The wealthiest Romans built extensive villa gardens with water features, topiary and cultivated roses and shaded arcades. Archeological evidence survives at sites such as Hadrian's Villa.

Byzantium and Moorish Spain kept garden traditions alive after the 4th century and the fall of Rome. By this time, a separate gardening tradition had arisen in China, which was transmitted to Japan, where it developed into aristocratic miniature landscapes centered on ponds and separately into the severe Zen gardens of temples.

In Europe, gardening revived in Languedoc and the Île-de-France in the 13th century. The rediscovery of descriptions of antique Roman villas and gardens led to the creation of a new form of garden, the Italian Renaissance garden in the late 15th and early 16th century. The first public parks were built by the Spanish Crown in the 16th century, in Europe and the Americas. The formal Garden à la française, exemplified by the Gardens of Versailles, became the dominant style of garden in Europe until the middle of the 18th century when it was replaced by the English landscape garden and the French landscape garden. The 19th century saw a welter of historical revivals and Romantic cottage-inspired gardening. In England, William Robinson and Gertrude Jekyll were strong proponents of the wild garden and the perennial garden respectively. Andrew Jackson Downing and Frederick Law Olmsted adapted European styles for North America, especially influencing public parks, campuses and suburban landscapes. Olmsted's influence extended well into the 20th century.

The 20th century saw the influence of modernism in the garden: from the articulate clarity of Thomas Church to the bold colors and forms of Brazilian Roberto Burle Marx.

A strong environmental consciousness and Sustainable design practices, such as green roofs and rainwater harvesting, are driving new considerations in gardening today.

The Historical Development of Garden Styles

Mesopotamian Gardens

Map showing the Tigris and Euphrates Rivers

Mesopotamia - the "land between the Rivers" Tigris and Euphrates - comprises a hilly and mountainous northern area and a flat, alluvial south. Its peoples (Sumerians, Akkadians, Assyrians, and Babylonians) were urban and literate from about 3,000BC. Evidence for their gardens comes from written texts, pictorial sculpture and archaeology. In western tradition Mesopotamia was the location of the Garden of Eden and the Hanging Gardens of Babylon. Temple gardens developed from the representation of a sacred grove; several distinct styles of royal garden are also known.

The courtyard garden was enclosed by the walls of a palace, or on a larger scale was a cultivated place inside the city walls. At Mari on the Middle Euphrates (c 1,800BC) one of the huge palace courtyards was called the Court of the Palms in contemporary written records. It is crossed by raised walkways of baked brick; the king and his entourage would dine there. At Ugarit (c1,400BC) there was a stone water basin, not located centrally as in later Persian gardens, for the central feature was probably a tree (date palm or tamarisk). The 7th century BC Assyrian king Assurbanipal is shown on a sculpture feasting with his queen, reclining on a couch beneath an arbour of vines, attended by musicians. Trophies of conquest are on display, including the dismembered head of the king of Elam hanging from a fragrant pine branch! A Babylonian text from the same period is divided into sections as if showing beds of soil with the names of medicinal, vegetable and herbal plants written into each square, perhaps representing a parterre design.

On a larger scale royal hunting parks were established to hold the exotic animals and plants which the king had acquired on his foreign campaigns. King Tiglath-Pileser I (c 1,000BC) lists horses, oxen, asses, deer of two types, gazelle and ibex, boasting "I numbered them like flocks of sheep."

From around 1,000 BC the Assyrian kings developed a style of city garden incorporating a naturalistic layout, running water supplied from river headwaters, and exotic plants from their foreign campaigns. Assurnasirpal II (883-859BC) lists pines of different kinds, cypresses and junipers of different kinds, almonds, dates, ebony, rosewood, olive, oak, tamarisk, walnut, terebinth and ash, fir pomegranate, pear, quince, fig and grapevines: "The canal water gushes from above into the gardens; fragrance pervades the walkways; streams of water as numerous as the stars of heaven flow in the pleasure garden.... Like a squirrel I pick fruit in the garden of delights." The city garden reached its zenith with the palace design of Sennacherib (704-681BC) whose water system stretched for 50 km into the hills, whose garden was higher and more ornate than any others, and who boasted of the complex technologies he deployed, calling his palace and garden "a Wonder for all Peoples".

The biblical Book of Genesis mentions the Tigris and Euphrates as two of the four rivers bounding the Garden of Eden. No specific place has been identified although there are many theories.

The Hanging Gardens of Babylon are listed by classical Greek writers as one of the Seven Wonders of the World - places to see before you die. The excavated ruins of Babylon do not reveal any suitable evidence, which has led some scholars to suggest that they may have been purely legendary. Alternatively the story may have originated from Sennacherib's garden in Nineveh.

Persian Gardens

The large charbagh (a form of Persian garden divided into four parts) provides the foreground for the classic view of the Taj Mahal, UNESCO World Heritage Site

All Persian gardens, from the ancient to the high classical were developed in opposition to the harsh and arid landscape of the Iranian Plateau. Unlike historical European gardens, which seemed carved or re-ordered from within their existing landscape, Persian gardens appeared as impossibilities. Their ethereal and delicate qualities emphasized their intrinsic contrast to the hostile environment. Trees and trellises largely feature as biotic shade; pavilions and walls are also structurally prominent in blocking the sun.

The heat also makes water important, both in the design and maintenance of the garden. Irrigation may be required, and may be provided via a form of underground tunnel called a qanat, that transports water from a local aquifer. Well-like structures then connect to the qanat, enabling the drawing of water. Alternatively, an animal-driven Persian well would draw water to the surface. Such wheel systems also moved water around surface water systems, such as those in the chahar bāgh style. Trees were often planted in a ditch called a juy, which prevented water evaporation and allowed the water quick access to the tree roots.

The Persian style often attempts to integrate indoors with outdoors through the connection of a surrounding garden with an inner courtyard. Designers often place architectural elements such as vaulted arches between the outer and interior areas to open up the divide between them.

Egyptian Gardens

Gardens were much cherished in the Egyptian times and were kept both for secular purposes and attached to temple compounds. Gardens in private homes and villas before the New Kingdom were mostly used for growing vegetables and located close to a canal or the river. However, in the New Kingdom they were often surrounded by walls and their purpose incorporated pleasure and beauty besides utility. Garden produce made out an important part of foodstuff but flowers were also cultivated for use in garlands to wear at festive occasions and for medicinal purposes. While the poor kept a patch for growing vegetables, the rich people could afford gardens lined with sheltering trees and decorative pools with fish and waterfowl. There could be wooden structures forming pergolas to support vines of grapes from which raisins and wine were produced. There could even be elaborate stone kiosks for ornamental reasons, with decorative statues.

Rectangular fishpond with ducks and lotus planted round with date palms and fruit trees, in a fresco from the Tomb of Nebamun, Thebes, 18th Dynasty

A funerary model of a garden, dating to the Eleventh dynasty of Egypt, c. 2009–1998 BC. Made of painted and gessoed wood, originally from Thebes.

Temple gardens had plots for cultivating special vegetables, plants or herbs considered sacred to a certain deity and which were required in rituals and offerings like lettuce to Min. Sacred groves and ornamental trees were planted in front of or near both cult temples and mortuary temples. As temples were representations of heaven and built as the actual home of the god, gardens were laid out according to the same principle. Avenues leading up to the entrance could be lined with trees, courtyards could hold small gardens and between temple buildings gardens with trees, vineyards, flowers and ponds were maintained.

The ancient Egyptian garden would have looked different from a modern garden. It would have seemed more like a collection of herbs or a patch of wild flowers, lacking the specially bred flowers of today. Flowers like the iris, chrysanthemum, lily and delphinium (blue), were certainly known to the ancients but do not feature much in garden scenes. Formal boquets seem to have been composed of mandrake, poppy, cornflower and or lotus and papyrus.

Due to the arid climate of Egypt, tending gardens meant constant attention and depended on irrigation. Skilled gardeners were employed by temples and households of the wealthy. Duties included planting, weeding, watering by means of a shaduf, pruning of fruit trees, digging the ground, harvesting the fruit etc.

Hellenistic and Roman Gardens

Hellenistic Gardens

It is curious that although the Egyptians and Romans both gardened with vigor, the Greeks did not own private gardens. They did put gardens around temples and they adorned walkways and roads with statues, but the ornate and pleasure gardens that demonstrated wealth in the other communities is seemingly absent.

Reconstruction of the Roman garden of the House of the Vettii in Pompeii

Roman Gardens

Roman gardens were a place of peace and solitude, a refuge from urban life. Ornamental horticulture became highly developed during the development of Roman civilization. The administrators of the Roman Empire (c.100 BC - 500 AD) actively exchanged information on agriculture, horticulture, animal husbandry, hydraulics, and botany. Seeds and plants were widely shared. The Gardens of Lucullus (Horti Lucullani) on the Pincian Hill on the edge of Rome introduced the Persian garden to Europe, about 60 BC.

Chinese and Japanese Gardens

Rock sculpture from the 'Lingering Garden' of Suzhou, China

Both Chinese and Japanese garden design traditionally is intended to evoke the natural landscape of mountains and rivers. However, the intended viewpoint of the gardens differs: Chinese gardens were intended to be viewed from within the garden and are intended as a setting for everyday life. Japanese gardens, with a few exceptions, were intended to be viewed from within the house, somewhat like a diorama. Additionally, Chinese gardens more often included a water feature, while Japanese gardens, set in a wetter climate, would often get by with the suggestion of water. (Such as sand or pebbles raked into a wave pattern.) Traditional Chinese gardens are also more likely to treat the plants in a naturalistic way, while traditional Japanese gardens might feature plants sheared into mountain shapes. This contrasts with the handling of stone elements: in a Japanese garden, stones are placed in groupings as part of the landscape, but in a Chinese garden, a particularly choice stone might even be placed on a pedestal in a prominent location so that it might be more easily appreciated.

Chinese Scholar Gardens

The style of Chinese garden varies among economic groups and differs by dynasties. Rocks, water, bridges and pavilions are among the most common features of scholar gardens for the wealthy classes, while courtyards, wells, and terra cotta fish tanks are common among general population. Other features such as moon gates and leaky windows (openwork screens that pierce surrounding walls) are seen in both groups.

The development of landscape design in China was historically driven by philosophies of both Confucianism and Taoism. Geometric symmetry and reinforcement of class boundaries were typical characteristics of landscape design in Asian cities, and both characteristics reflect Confucian ideals. While the British used nature outside the home to provide privacy, Chinese homes were compounds made of a number of buildings which all faced one or more courtyards or common areas. Rather than around the home, the Chinese valued natural spaces inside the compound, which is where the family socialized. Furthermore, Courtyards in the Chinese home reflected Taoist philosophies, where families would try to create abstractions of nature rather than recreations of it. For example, a Taoist garden would avoid straight lines and use stone and water instead of trees, whereas Asian cities followed Confucian, geometric designs and North American parks typically feature trees and lawns.

There are two ways of looking at the signature design characteristics of the Chinese garden: first, the concept of Yin and Yang and second, the myths of longevity that arose during the Qin Dynasty.

The philosophy of Yin and Yang portrays the idea of balance and harmony. The Chinese garden expresses the relationship to nature and the idea of balance through the art of mimicking natural setting, thus the existence of mountains, rocks, water, and wind elements. Yin and Yang juxtapose complementary opposites: as hard as rock can be, the softness of water can dissolve it. Lake Tai rocks, limestone eroded by the water of Lake

Tai, are the quintessential example. Water, air and light run through the rock as it sit still on display. The leaky windows of the Chinese garden wall portray both steadiness and movement. The windows create a solid painting on walls, however that steadiness changes once the wind blows or the eyes move.

Chinese garden's structure is based upon the culture's creation myth, rooted in rocks and water. To have longevity is to live among mountains and water; it is to live with nature, to live like an immortal being (Xian). The garden evokes a healthy lifestyle that makes one immortal, free from the problems of civilization. Thus, Chinese landscape is known as Shan (mountain) and Shui (water). (Add Roger's citation).

Symbolism is a key element of Chinese garden design. To the earthy tones of the Chinese garden, a touch of red or gold is often added to bring forth the Yin/Yang contrast. The colors red and gold also represent luck and wealth. Bats, dragons and other mystic creatures carved on wooden doors are also commonly found in Chinese gardens; these are signs of luck and protection.

Circles portray togetherness, especially for family members, and are depicted in moon gates and round tables placed within square backgrounds. The moon gate and other whimsical doorways also act to frame views and to force the viewer to pause for a transition into a new space.

Paths in Chinese gardens are often uneven and sometimes consciously zigzag. These paths are like the passages of a human life. There is always something new or different when seen from a different angle, while the future is unknown and unpredictable.

European Gardens

Gardens of Byzantium

The Byzantine empire span a period of more than 1000 years (330-1453 AD) and a geographic area from modern day Spain and Britain to the Middle East and north Africa. Probably due to this temporal and geographic spread and its turbulent history, there is no single dominant garden style that can be labeled "Byzantine style". Archaeological evidence of public, imperial, and private gardens is scant at best and researchers over the years have relied on literary sources to derive clues about the main features of Byzantine gardens. Romance novels such as Hysmine and Hysminias (12th century) included detailed descriptions of gardens and their popularity attests to the Byzantines' enthusiasm for pleasure gardens (locus amoenus). More formal gardening texts such as the Geoponika (10th century) were in fact encyclopedias of accumulated agricultural practices (grafting, watering) and pagan lore (astrology, plant sympathy/antipathy relationships) going back to Hesiod's time. Their repeated publications and translations to other languages well into the 16th century is evidence to the value attributed to the horticultural knowledge of antiquity. These literary sources worked as handbooks promoting the concepts of walled gardens with plants arranged by type. Such ideals found expression in the suburban parks (Philopation, Aretai) and palatial gardens (Mesoke-pion, Mangana) of Constantinople.

The Byzantine garden tradition was influenced by the strong undercurrents of history that the empire itself was exposed to. The first and foremost influence was the adoption of Christianity as the empire's official religion by its founder Constantine I. The new religion signaled a departure from the ornamental pagan sculptures of the Greco-Ro-man garden style. The second influence was the increasing contact with the Islamic nations of the Middle East especially after the 9th century. Lavish furnishings in the emperor's palace and the adoption of automata in the palatial gardens are evidence of this influence. The third factor was a fundamental shift in the design of the Byzantine cities after the 7th century when they became smaller in size and population as well as more ruralized. The class of wealthy aristocrats who could finance and maintain elabo-rate gardens probably shrank as well. The final factor was a shifting view toward a more "enclosed" garden space (hortus conclusus); a trend dominant in Europe at that time. The open views and vistas so much favored by the garden builders of the Roman villas were replaced by garden walls and scenic views painted on the inside of these walls. The concept of the heavenly paradise was an enclosed garden gained popularity during that time and especially after the iconoclastic period (7th century) with the emphasis it placed on divine punishment and repentance.

An area of horticulture that flourished throughout the long history of Byzantium was that practiced by monasteries. Although archaeological evidence has provided limited evidence of monastic horticulture, a great deal can be learned by studying the founda-tion documents of the numerous Christian monasteries as well as the biographies of saints describing their gardening activities. From these sources we learn that monas-teries maintained gardens outside their walls and watered them with complex irriga-tion systems fed by springs or rainwater. These gardens contained vineyards, broadleaf vegetables, and fruit trees for the sustenance of monks and pilgrims alike. The role of the gardener was frequently assumed by monks as an act of humility. Monastic hor-ticultural practices established at that time are still in use in Christian monasteries throughout Greece and the Middle East.

Medieval

Monasteries carried on a tradition of garden design and intense horticultural tech-niques during the medieval period in Europe. Rather than any one particular horticul-tural technique employed, it is the variety of different purposes the monasteries had for their gardens that serves as testament to their sophistication. As for gardening prac-tices, records are limited, and there are no extant monastic gardens that are entirely true to original form. There are, however, records and plans that indicate the types of garden a monastery might have had, such as those for St. Gall in Switzerland.

Generally, monastic garden types consisted of kitchen gardens, infirmary gardens, cemetery orchards, cloister garths and vineyards. Individual monasteries might also have had a "green court", a plot of grass and trees where horses could graze, as well as a cellarer's garden or private gardens for obedientiaries, monks who held specific posts within the monastery.

From a utilitarian standpoint, vegetable and herb gardens helped provide both alimentary and medicinal crops, which could be used to feed or treat the monks and, in some cases, the outside community. As detailed in the plans for St. Gall, these gardens were laid out in rectangular plots, with narrow paths between them to facilitate collection of yields. Often these beds were surrounded with wattle fencing to prevent animals from entry. In the kitchen gardens, fennel, cabbage, onion, garlic, leeks, radishes, and parsnips might be grown, as well as peas, lentils and beans if space allowed for them. The infirmary gardens could contain Rosa gallica ("The Apothecary Rose"), savory, costmary, fenugreek, rosemary, peppermint, rue, iris, sage, bergamot, mint, lovage, fennel and cumin, amongst other herbs. From a utilitarian standpoint, vegetable and herb gardens helped provide both alimentary and medicinal crops, which could be used to feed or treat the monks and, in some cases, the outside community.

The herb and vegetable gardens served a purpose beyond that of production, and that was that their installation and maintenance allowed the monks to fulfill the manual labor component of the religious way of life prescribed by Rule of St. Benedict.

Orchards also served as sites for food production and as arenas for manual labor, and cemetery orchards, such as that detailed in the plan for St. Gall, showed yet more versatility. The cemetery orchard not only produced fruit, but manifested as a natural symbol of the garden of Paradise. This bi-fold concept of the garden as a space that met both physical and spiritual needs was carried over to the cloister garth.

The cloister garth, a claustrum consisting of the viridarium, a rectangular plot of grass surrounded by peristyle arcades, was barred to the laity, and served primarily as a place of retreat, a locus of the 'vita contempliva'. The viridarium was often bisected or quartered by paths, and often featured a roofed fountain at the center or side of the garth that served as a primary source for wash water and for irrigation, meeting yet more physical needs. Some cloister gardens contained small fish ponds as well, another source of food for the community. The arcades were used for teaching, sitting and meditating, or for exercise in inclement weather.

There is much conjecture as to ways in which the garth served as a spiritual aid. Umberto Eco describes the green swath as a sort of balm on which a monk might rest weary eyes, so as to return to reading with renewed vigor. Some scholars suggest that, though sparsely planted, plant materials found in the cloister garth might have inspired various religious visions. This tendency to imbue the garden with symbolic values was not inherent to the religious orders alone, but was a feature of medieval culture in general. The square closter garth was meant to represent the four points of the compass, and so the universe as a whole. As Turner puts it,

Augustine inspired medieval garden makers to abjure earthliness and look upward for divine inspiration. A perfect square with a round pool and a pentagonal fountain became a microcosm, illuminating the mathematical order and divine grace of the macrocosm (the universe).

Walking around the cloister while meditating was a way of devoting oneself to the "path of life"; indeed, each of the monastic gardens was imbued with symbolic as well as palpable value, testifying to the ingenuity of its creators.

In the later Middle Ages, texts, art and literary works provide a picture of developments in garden design. During the late 12th through 15th centuries, European cities were walled for internal defense and to control trade. Though space within these walls was limited, surviving documents show that there were animals, fruit trees and kitchen gardens inside the city limits.

Pietro Crescenzi, a Bolognese lawyer, wrote twelve volumes on the practical aspects of farming in the 13th century and they offer a description of medieval gardening practices. From his text we know that gardens were surrounded with stonewalls, thick hedging or fencing and incorporated trellises and arbors. They borrowed their form from the square or rectangular shape of the cloister and included square planting beds.

Grass was also first noted in the medieval garden. In the De Vegetabilibus of Albertus Magnus written around 1260, instructions are given for planting grass plots. Raised banks covered in turf called "Turf Seats" were constructed to provide seating in the garden. Fruit trees were prevalent and often grafted to produce new varieties of fruit. Gardens included a raised mound or mount to serve as a stage for viewing and planting beds were customarily elevated on raised platforms.

Two works from the late Middle Ages discuss plant cultivation. In the English poem "The Feate of Gardinage" by Jon Gardener and the general household advice given in Le Ménagier de Paris of 1393, a variety of herbs, flowers, fruit trees and bushes were listed with instructions on their cultivation. The Menagier provides advice by season on sowing, planting and grafting. The most sophisticated gardening during the Middle Ages was done at the monasteries. Monks developed horticultural techniques, and cultivated herbs, fruits and vegetables. Using the medicinal herbs they grew, monks treated those suffering inside the monastery and in surrounding communities.

During the Middle Ages, gardens were thought to unite the earthly with the divine. The enclosed garden as an allegory for paradise or a "lost Eden" was termed the Hortus Conclusus. Freighted with religious and spiritual significance, enclosed gardens were often depicted in the visual arts, picturing the Virgin Mary, a fountain, a unicorn and roses inside an enclosed area.

Though Medieval gardens lacked many of the features of the Renaissance gardens that followed them, some of the characteristics of these gardens continue to be incorporated today.

The Renaissance

- The Italian Renaissance inspired a revolution in private gardening. Renaissance private gardens were full of scenes from ancient mythology and other learned allusions. Water during this time was especially symbolic: it was associated with fertility and the abundance of nature.

- The first public gardens were built by the Spanish Crown in the 16th century, in Europe and the Americas.

- Terraced Italian Renaissance gardens

The Medici Villa Petraia, near Florence, laid out by Niccolò Tribolo, epitomizes the Italian garden of the early Renaissance, before the grander architectural schemes of the 16th century

French Baroque

Portrait of André Le Nôtre (12 March 1613-15 September 1700) by Carlo Maratta

- The Garden à la française, or Baroque French gardens, in the tradition of André Le Nôtre.

The French Classical garden style, or Garden à la française, climaxed during the reign of Louis XIV of France (1638–1715) and his head gardener of Gardens of Versailles, André Le Nôtre (1613–1700). The inspiration for these gardens initially came from the

Italian Renaissance garden of the 14th and 15th centuries and ideas of French philosopher René Descartes (1576–1650). At this time the French opened the garden up to enormous proportions compared to their Italian predecessor. Their gardens epitomize monarch and 'man' dominating and manipulating nature to show his authority, wealth, and power.

Renée Descartes, the founder of analytical geometry, believed that the natural world was objectively measurable and that space is infinitely divisible. His belief that "all movement is a straight line therefore space is a universal grid of mathematical coordinates and everything can be located on its infinitely extendable planes" gave us Cartesian mathematics. Through the classical French gardens this coordinate system and philosophy is now given a physical and visual representation.

This French formal and axial garden style placed the house centrally on an enormous and mainly flat property of land. A large central axis that gets narrower further from the main house, forces the viewer's perspective to the horizon line, making the property look even larger. The viewer is to see the property as a cohesive whole but at the same time is unable to see all the components of the garden. One is to be led through a logical progression or story and be surprised by elements that aren't visible until approached. There is an allegorical story referring to the owner through statues and water features which have mythological references. There are small, almost imperceptible grade changes that help conceal the gardens surprises as well as elongate the gardens views.

These grand gardens have organized spaces meant to be elaborate stages for entertaining the court and guests with plays, concerts and fireworks displays. The following list of garden features were used:

- Allée
- Axis
- Bosquet
- Canal
- Cul de sac
- Fountains
- Grottos with rocaille
- Orangerie
- Parterre de broderie
- Patte d'oie (Goose foot)
- Tapis Vert
- Topiary

Mediterranean Gardens

Due to being an early hub for Western society and being used for centuries, Mediter-ranean soil was fragile, and one could think of the region's landscape culture to be a conflict between fruitfulness and frugality. The area consisted largely of small-scale agricultural plots. Later, following World War II, Mediterranean immigrants brought this agricultural style to Canada, where fruit trees and vegetables in the backyard be-came common.

Anglo-Dutch Gardens

• Anglo-Dutch formal gardens

Picturesque and English Landscape Gardens

Forested areas played a number of roles for the British in the Middle Ages, and one of those roles was to produce game for the gentry. Lords of valuable land were expected to provide a bounty of animals for hunting during royal visits. Despite being in natu-ral locations, forested manor homes could symbolize status, wealth and power if they appeared to have all amenities. After the Industrial Revolution, Britain's forest indus-try shrank until it no longer existed. In response, the Garden City Movement brought urban planning into industrialized areas in the early 20th century to offset negative industrial effects such as pollution.

There were several traditions that influenced English gardening in the 18th century, the first of which was to plant woods around homes. By the mid-17th century, cop-pice planting became consistent and was considered visually and aesthetically pleasing. Whereas forested areas were more useful for hunting purposes in Britain during the Middle Ages, 18th century patterns demonstrate a further deviation in gardening ap-proach from practicality toward design meant to please the senses.

Likewise, English pleasure grounds were influenced by medieval groves, some of which were still in existence in 18th century Britain. This influence manifest in the form of shrubbery, sometimes organized in mazes or maze-like formations. And though also ancient, shredding became a common characteristic of these early gardens, as the method enabled light to enter the understory. Shredding was used to make garden groves, which ideally included an orchard with fruit trees, fragrant herbs and flowers, and moss-covered pathways.

The picturesque garden style emerged in England in the 18th century, one of the grow-ing currents of the larger Romantic movement. Garden designers like William Kent and Capability Brown emulated the allegorical landscape paintings of European artists, especially Claude Lorraine, Poussin and Salvator Rosa. The manicured hills, lakes and trees dotted with allegorical temples were sculpted into the land.

By the 1790s there was a reaction against these stereotypical compositions; a number of thinkers began to promote the idea of picturesque gardens. The leader of the movement was landscape theorist William Gilpin, an accomplished artist known for his realistic depictions of Nature. He preferred the natural landscape over the manicured and urged designers to respond to the topography of a given site. He also noted that while classical beauty was associated with the smooth and neat, picturesque beauty had a wilder, untamed quality. The picturesque style also incorporated architectural follies—castles, Gothic ruins, rustic cottages—built to add interest and depth to the landscape

Controversy between the picturesque school and proponents of the more manicured garden raged well into the 19th century. Landscape designer Humphry Repton supported Gilpin's ideas, particularly that of the garden harmonizing with surrounding landforms. He was attacked in the press by two rival theorists, Richard Payne Knight and Uvedale Price. Repton countered by highlighting the differences between painting and landscape gardening, William Shenstone has been credited with coining the term 'landscape gardening'. Unlike a painting, the viewer moves through a garden, constantly shifting viewpoints.

The French landscape garden, also called the jardin anglais or jardin pittoresque, was influenced by contemporary English gardens. Rococo features like Turkish tents and Chinese bridges are prevalent in French gardens in the 18th century. The French Picturesque garden style falls into two categories: those that were staged, almost like theatrical scenery, usually rustic and exotic, called jardin anglo-chinois, and those filled with pastoral romance and bucolic sentiment, influenced by Jean-Jacques Rousseau. The former style is represented by the Désert de Retz and Parc Monceau, the latter by the Moulin Jolie.

The rusticity found in French picturesque gardens is also derived from admiration of Dutch 17th-century landscape painting and works of the French 18th-century artists Claude-Henri Watelet, François Boucher and Hubert Robert.

English garden is the common term in the English-speaking world for interpretations, derivations, and revivals in the style of the original Landscape Garden examples.

"Gardenesque" Gardens

The "Gardenesque" style of English garden design evolved during the 1820s from Humphry Repton's Picturesque or "Mixed" style, largely through the efforts of J. C. Loudon, who invented the term.

In a Gardenesque plan, all trees, shrubs and other plants are positioned and managed in such a way that the character of each plant can be displayed to its full potential. With the spread of botany as a suitable subject of study for the enlightened, the Gardenesque tended to emphasize botanical curiosities and a collector's approach. New plant material that would have seemed bizarre and alien in earlier gardening found settings: pampas grass from Argentina and monkey-puzzle trees from Chile, for example. Winding paths linked scattered plantings. The Gardenesque approach involved the creation of small-scale landscapes, dotted with features and vignettes, to promote beauty of detail, variety and mystery, sometimes to the detriment of coherence. Artificial mounds helped to stage groupings of shrubs, and island beds became prominent features.

"Wild" Gardens and Herbaceous Borders

The books of William Robinson describing his own "wild" gardening at Gravetye Manor in Sussex, and the sentimental picture of a rosy, idealized "cottage garden" of the kind pictured by Kate Greenaway, which had scarcely existed historically, both influenced the development of the mixed herbaceous borders that were advocated by Gertrude Jekyll at Munstead Wood in Surrey from the 1890s. Her plantings, which mixed shrubs with perennial and annual plants and bulbs in deep beds within more formal structures of terraces and stairs designed by Edwin Lutyens, set the model for high-style, high-maintenance gardening until the Second World War. Vita Sackville-West's garden at Sissinghurst Castle, Kent is the most famous and influential garden of this last blossoming of romantic style, publicized by the gardener's own gardening column in The Observer. The trend continued in the gardening of Margery Fish at East Lambrook Manor. In the last quarter of the 20th century, less structured Wildlife gardening emphasized the ecological framework of similar gardens using native plants. A leading proponent in the United States was the landscape architect Jens Jensen. He designed city and regional parks, and private estates, with a honed aesthetic of art and nature.

Pattern Gardens: Revived Parterres

Contemporary Gardens

- Romantic English cottage garden revival.

- Modernist gardens.

- Naturalistic habitat gardens

In the 20th century, modern design for gardens became important as architects began to design buildings and residences with an eye toward innovation and streamlining the formal Beaux-Arts and derivative early revival styles, removing unnecessary references and embellishment. Garden design, inspired by modern architecture, naturally followed in the same philosophy of "form following function". Thus concerning the many philosophies of plant maturity. In post-war United States people's residences and domestic lives became more outdoor oriented, especially in the western states as promoted by 'Sunset Magazine', with the backyard often becoming an outdoor room.

Frank Lloyd Wright demonstrated his interpretation for the modern garden by designing homes in complete harmony with natural surroundings. Taliesin and Fallingwater are both examples of careful placement of architecture in nature so the relationship between the residence and surroundings become seamless. His son Lloyd Wright trained in architecture and landscape architecture in the Olmsted Brothers office, with his father, and with architect Irving Gill. He practiced an innovative organic integration of structure and landscape in his works.

Subsequently Garrett Eckbo, James Rose, and Dan Kiley - known as the "bad boys of Harvard", met while studying traditional landscape architecture became notable pioneers in the design of modern gardens. As Harvard embraced modern design in their school of architecture, these designers wanted to interpret and incorporate those new ideas in landscape design. They became interested in developing functional space for outdoor living with designs echoing natural surroundings. Modern gardens feature a fresh mix of curved and architectonic designs and many include abstract art in geometrics and sculpture. Spaces are defined with the thoughtful placement of trees and plantings. Thomas Church work in California was influential through his books and other publications. In Sonoma County, California his 1948 Donnell garden's swimming Pool, kidney-shaped with an abstract sculpture within it, became an icon of modern outdoor living.

In Mexico Luis Barragán explored a synthesis of International style modernism with native Mexican tradition. in private estates and residential development projects such as Jardines del Pedregal (English: Rocky Gardens) and the San Cristobal 'Los Clubes' Estates in Mexico City. In civic design the Torres de Satélite are urban sculptures of substantial dimensions in Naucalpan, Mexico. His house, studio, and gardens, built in 1948 in Mexico City, was listed as a UNESCO World Heritage site in 2004.

Roberto Burle Marx is accredited with having introduced modernist landscape architecture to Brazil. He was known as a modern nature artist and a public urban space designer. He was landscape architect (as well as a botanist, painter, print maker, ecologist, naturalist, artist, and musician) who designed of parks and gardens in Brazil, Argentina, Venezuela, Kuala Lumpur, Malaysia, and in the United States in Florida. He worked with the architects Lúcio Costa and Oscar Niemeyer on the landscape design for some of the prominent modernist government buildings in Brazil's capitol Brasília.

References

- Cross, Rob; Spencer, Roger (2009). Sustainable Gardens. Collingwood, Australia: CSIRO. ISBN 9780643094222.

- Carroll, Steven B.; Salt, Steven B. (2004). Ecology for Gardeners. Portland, USA and Cambridge, UK: Timber Press. ISBN 0881926116.

- Scherr, Sara J.; Sthapit, Sajal (2009). Mitigating Climate Change through Food and Land Use (PDF). Washington, United States of America: Worldwatch Institute. ISBN 9781878071910.

- Allwood, Julian; Cullen, Jonathan (2011). Sustainable Materials - with both eyes open. Cambridge: UIT. ISBN 9781906860059.

- Reay, Dave; Smith, Pete; van Amstel, Andre (2010). Methane and Climate Change. London: Earthscan. ISBN 978-1844078233.

- Harriet Kopinska; Jane Griffiths; Heather Jackson; Pauline Pears (2011). The Garden Organic Book of Compost. London: New Holland. ISBN 9781847734372.

- Pond Conservation (2011). Creating a Garden Pond for Wildlife. Oxford: Freshwater Habitats Trust. ISBN 978-0-9537971-2-7.

- Sutton, Mark; Reis, Stefan (2011). The nitrogen cycle and its influence on the European greenhouse gas balance. Centre for Ecology and Hydrology. ISBN 978-1-906698-21-8.

- Morison, James I. L.; Morecroft, Michael D. (2006). Plant Growth and Climate Change. Oxford: Blackwell Publishing. ISBN 978-14051-3192-6.

- Broadmeadow, Mark; Ray, Duncan (2005). Climate Change and British Woodland (PDF). Edinburgh: Forestry Commission. ISBN 0-85538-658-4.

- Green, Charlotte (1999). Gardening Without Water: Creating beautiful gardens using only rainwater. Tunbridge Wells: Search Press. ISBN 0855328851.

- Dunnett, Nigel; Clayden, Andy (2007). Rain Gardens: Managing Water Sustainably in the Garden and Designed Landscape. Portland, Oregon, USA: Timber Press. ISBN 978-0881928266.

- Wilson, Matthew (2007). New Gardening: How to garden in a changing climate. London: Mitchell Beazley and the Royal Horticultural Society. ISBN 9781845333058.

Permissions

All chapters in this book are published with permission under the Creative Commons Attribution Share Alike License or equivalent. Every chapter published in this book has been scrutinized by our experts. Their significance has been extensively debated. The topics covered herein carry significant information for a comprehensive understanding. They may even be implemented as practical applications or may be referred to as a beginning point for further studies.

We would like to thank the editorial team for lending their expertise to make the book truly unique. They have played a crucial role in the development of this book. Without their invaluable contributions this book wouldn't have been possible. They have made vital efforts to compile up to date information on the varied aspects of this subject to make this book a valuable addition to the collection of many professionals and students.

This book was conceptualized with the vision of imparting up-to-date and integrated information in this field. To ensure the same, a matchless editorial board was set up. Every individual on the board went through rigorous rounds of assessment to prove their worth. After which they invested a large part of their time researching and compiling the most relevant data for our readers.

The editorial board has been involved in producing this book since its inception. They have spent rigorous hours researching and exploring the diverse topics which have resulted in the successful publishing of this book. They have passed on their knowledge of decades through this book. To expedite this challenging task, the publisher supported the team at every step. A small team of assistant editors was also appointed to further simplify the editing procedure and attain best results for the readers.

Apart from the editorial board, the designing team has also invested a significant amount of their time in understanding the subject and creating the most relevant covers. They scrutinized every image to scout for the most suitable representation of the subject and create an appropriate cover for the book.

The publishing team has been an ardent support to the editorial, designing and production team. Their endless efforts to recruit the best for this project, has resulted in the accomplishment of this book. They are a veteran in the field of academics and their pool of knowledge is as vast as their experience in printing. Their expertise and guidance has proved useful at every step. Their uncompromising quality standards have made this book an exceptional effort. Their encouragement from time to time has been an inspiration for everyone.

The publisher and the editorial board hope that this book will prove to be a valuable piece of knowledge for students, practitioners and scholars across the globe.

Index

www.ingramcontent.com/pod-product-compliance
Lightning Source LLC
Chambersburg PA
CBHW061931190326
41458CB00009B/2717